Praise for
How Women

Named a Summer Reading ...
Oprah.com and *Savoir Flair*

"Using a wealth of economic and social science research, Huston, a cognitive psychologist, . . . documents these stereotypes and shows how women are often trapped in situations where they can't come out ahead, no matter what they do . . . [*How Women Decide*] will resonate with any women trying to navigate treacherous career waters as well as with managers wondering how to increase diversity and get the best out of all their employees. One could also imagine it becoming required reading on Wall Street, where male-dominated thinking has caused so many problems." —*New York Times Book Review*

"Huston, writing in a cheerful, classroom voice, wants to give readers tools to take apart the frequently hostile response to women's decisions . . . In clear, declarative prose, [*How Women Decide*] dips readers' toes into stereotype threat and confirmation bias, role congruity theory, cortisol, and stress studies and prospect theory." —*Seattle Times*

"If you're a woman, read it. If you're a man, read it . . . Sometimes a book tells you something you really needed to know, whether you realised it or not, and *How Women Decide* is one of those books . . . [Huston] throws a bright laboratory light on familiar territory—women's experiences at work—and then pins down with scientific precision the subtle and not-so-subtle stereotyping women encounter, explaining how these societal expectations impact on women's decision-making." —*The National*

"A journey to break down barriers and open the conversation on how to shape habits, perceptions, and strategies to transform our society as a whole, regardless of who's making the decisions."

— *Savoir Flair*

"If you want to get deep into the ways we are swayed to make decisions that favor what someone else wants, I recommend [*How Women Decide*] . . . Huston gives a persuasive argument that intuition isn't an exclusive tool of women."   — MsCareerGirl.com

"*How Women Decide* blows up several myths about female decision-making that everyone believes, women included. Through thoughtful analysis and lively, entertaining anecdotes, it teaches us what's really happening—how bias works. Every woman needs to read this well-researched and wonderfully reported book. She'll gain confidence through useful tactics for even better decision-making. Men should read it, too; they'll learn tactics that make women great leaders!"   —Joanna Barsh, best-selling author of
How Remarkable Women Lead and *Centered Leadership*

"Ever wonder whether 'women's instinct' is a real thing? Ever consider multiple points of view, only to be called 'wishy-washy'? In this brilliantly researched and entertaining book, Therese Huston reveals the ways in which understanding ourselves and thinking critically about gender biases can help us all make better choices. I'm already using it to strategize at work, and I predict that every reader will learn something new and useful in its pages."

—Jessica Bacal, editor of *Mistakes I Made at Work: 25 Influential Women Reflect on What They Got Out of Getting It Wrong* and director of the Wurtele Center for Work & Life at Smith College

"Finally! A well-researched book that affirms the fact that, despite their self-doubts, women make *great* decision-makers. This book will help you to compete with your male counterparts with courage and confidence." —Lois P. Frankel, Ph.D., author of
Nice Girls Don't Get the Corner Office and *See Jane Lead*

"How do women make decisions? In this thoughtful, well-researched book, Huston avoids pop-psych answers that assume all women are the same. Exploding stereotypes, but showing their effect on women's behavior, she offers intelligent guidance to the challenges and process of making decisions." —Carol Tavris, Ph.D., coauthor of
Mistakes Were Made (But Not by Me)

"None of the myriad decision-making bestsellers consider how their advice should differ for men and women. *How Women Decide* overthrows such one-sex-fits-all recommendations. It combines engaging stories and compelling research to reveal how our beliefs about men and women drive the way they make choices."
—Daniel Simons, Ph.D., coauthor of
The Invisible Gorilla: How Our Intuitions Deceive Us

"With verve, charm, and a ruthless reliance on data, [Huston] challenge[s] and ultimately disprove[s] several common assumptions about how women make decisions . . . Huston provides sharp observations, handy chapter summaries, and practical advice . . . She builds a convincing case that if businesses, government, and other organizations want to improve their decision-making at the highest levels, they need to have more women in the boardroom; and she provides women readers with concrete strategies to defuse existing stereotypes." —*Publishers Weekly*

"Extraordinarily readable—and a profound supplement to Sandberg's *Lean In*."   —*Booklist*

"Insightful advice for women about decisiveness, confidence, and tackling gender bias . . . Useful, practical strategies based on informed analysis."   —*Kirkus Reviews*

"An authoritative guide to help women navigate the workplace and their everyday life with greater success and impact."   —*Forbes*

"Contains advice for everyone."   —*Financial Times*

How Women Decide

Therese Huston

MARINER BOOKS
HOUGHTON MIFFLIN HARCOURT
BOSTON   NEW YORK

First Mariner Books edition 2017

Library of Congress Cataloging-in-Publication Data
Names: Huston, Therese, author.
Title: How women decide / Therese Huston.
Description: Boston : Houghton Mifflin Harcourt, 2016. |
Includes bibliographical references.
Identifiers: LCCN 2015037243 | ISBN 9780544416093 (hardcover) |
ISBN 9780544416109 (ebook) | ISBN 9780544944817 (pbk.)
Subjects: LCSH: Decision making — Sex differences. |
Decision making — Psychological aspects. | Women — Psychology.
Classification: LCC BF448 .H87 2016 | DDC 155.3/33 — dc23
LC record available at http://lccn.loc.gov/2015037243

Printed in the United States of America
DOC 10 9 8 7 6 5 4 3 2 1

To Jonathan,
because marrying you was the best decision I ever made

Contents

What Happens When a Woman
Makes the Call?

FROM EVERY DIRECTION LATELY, women are hearing a call to arms. Women have been told to lean in, ask for what they want, know their value, play big, don their bossypants, and close the confidence gap. These messages galvanize. They embolden women to take their proper seats at the table and they promise power to those who want it. If women work hard and raise their expectations, they're told, they will achieve the highest levels of success—and that means they will be making more of the big decisions.

But no one has talked about what happens to women when they make these big decisions. Is a woman's experience issuing a tough call, a decision with serious stakes, any different from a man's? That's the question that ignited my research and eventually caught fire as this book. I've found that when a man faces a hard decision, he only has to think about making a judgment, but when a woman faces a hard decision, she has to think about making a judgment and also navigate being judged.

What's a smart, self-respecting, and (let's face it) busy woman to do?

She needs to know how women decide and how to take the realities of the decision-making landscape into account when planning her own course of action. I'll share a secret with you: Women approach decisions in ways that are actually stronger than they realize. Men and women approach decisions differently, but not necessarily in the ways people have been led to believe. This isn't a "biology is destiny" or a pink brain / blue brain book. Society has been underestimating women's abilities to make astute choices for years, and this doubting, this routine questioning of a woman's judgment, drives many of the gender differences we see.

Often we don't realize that we're scrutinizing a woman's decision more than we would a man's; it can be hard to notice because there are very few scenarios where all factors other than gender are identical. Sometimes, though, a situation arises where we can see a clear parallel and a clear bias. Take, for example, the moment in February 2013 when Marissa Mayer made headlines for changing Yahoo's work-from-home policy. Yahoo announced that employees could no longer telecommute full-time, and the press lambasted Mayer. Pundits criticized the policy change, saying it would hurt women, and many of us, myself included, raised eyebrows about Mayer's controversial decision. But how many people heard about it when Best Buy's CEO, Hubert Joly, made the same decision about a week later?[1] When he ended Best Buy's generous work-from-home policy, business reporters dutifully picked up the story, but his announcement didn't cause a public outcry the way Mayer's did. Joly popped up in headlines for his decision briefly in 2013, but as late as 2015, journalists were still talking about Mayer's decision, analyzing whether she made the right choice.[2] So for making the same judgment call, a male CEO drew some sidelong glances for a few months, but a female CEO drew extensive scrutiny and censure for years.

At first, we tend to rationalize our reactions. Yahoo's decision must have impinged on more employees' schedules because it's a software company, and programmers can work in their pajamas at home at any hour of the day or night; Best Buy has stores, we reason, and employees need to appear fully clothed and on time. Their telecommuting pool must be tiny. But articles on the story indicated that Mayer's decision affected only two hundred employees, whereas Joly's decision reportedly changed the lives of nearly four thousand corporate employees who often worked from home.[3] That's twenty times more workers touched by the Best Buy decision.

If the number of affected employees doesn't explain the outcry against Mayer and the complacency around Joly, what does? Had Mayer just taken the helm at Yahoo while Joly was a fixture at Best Buy? No. This is where the parallels become even more unsettling — both chiefs had been on the job roughly six months.[4] One likely reason we keep fuming over Mayer's decision but ignore Joly's choice lies in a pattern that many of us unknowingly fall into: we're quick to question a woman's decision but inclined to accept a man's. Men and women don't have to act differently for us to see them differently.

This tendency has very real consequences. Consider the often-cited observation that businesses are eager to promote men but reluctant to promote women. Why? Your bookshelf may be full of answers to that question, but my research suggests a new one, one many people have overlooked. We trust men to make the hard choices. We are quick to accept a man's decisions, even the hard, unpleasant ones, as being what must be done. When a woman announces the same difficult decision, we scrutinize it with twice the vigor. We may not mean to, but we doubt the quality of her choices.

It may be hard to believe that decision-making has a gender component, that someone would give a man an understanding nod but give a woman a raised eyebrow for making the same call. We see ourselves as fair people with the best of intentions. I've never met a single person who has said, "I love to discriminate." If we want to understand how gender changes the decision-making process as well as the subtle and not-so-subtle ways we react to men's and women's choices, we need to ask some rigorous questions. Is there any real difference between men's and women's judgment? Might we ever exaggerate the gap? Where has popular culture exposed real disparities in the ways men and women decide, and where has popular culture actually manufactured the differences? In cases where women and men do take different approaches to the same choice, is the way women reach a decision ever an asset rather than a liability?

Most important, if we do find that there are differences in how men's and women's decisions are received, what can we do about it? How do we become more aware of our favoritism and catch ourselves in the act? Partly, we need to educate ourselves about our hidden biases around decision-making. Both men and women must take stock and strategize, because no one person can do this alone. Certainly, reading this book can and should help improve the decisions you make regardless of your gender, but if we want to see more women take meaningful seats at the table, we ought to change how we, as a culture, talk about women's judgment. We need to make some structural changes, and these changes will improve not just the lives of women but the decisions being made for our world. If you gain only one insight from this book, I hope it's this: Having a greater number of women in the room when a crucial decision is being made is not only better for women, it's better for the decision. And that's better for everyone.

Whom Do We Ask to Make Decisions of Consequence?

It was January 1968 and a typical winter day in Seattle, cold enough to make you bundle up but not cold enough to snow. Barbara Winslow was twenty-three years old and a history major at the University of

Washington, and she and her husband of less than a year were sitting in a doctor's office, not liking what they heard.

A few days earlier, Barbara had found a lump in her breast. The doctor explained he would sedate Barbara, take a slice out of her breast, run some tests while she was still under anesthesia, and, if the biopsy came back positive and the tissue was malignant, he would immediately perform a radical mastectomy. A radical mastectomy is aptly named. It entails removing the entire breast, the chest muscles underneath, and all the lymph nodes from a woman's underarm in one single, efficient, and slightly barbaric procedure.[5] Barbara would fall asleep wondering if she had cancer, and she would wake up to either fantastic news or stitches where her breast had been.

After describing the operation, the doctor said they should schedule the biopsy immediately. Barbara protested. Could she go home and think about it? Why did this decision have to be made right then? The doctor explained that if he gave her a lot of time to think about this radical procedure, she would likely become too frightened and decide not to take the risk.

Barbara had been taught that you didn't question the authority of doctors, so she said: "Okay, that makes sense." "Good," said the doctor, and he handed the consent form to her husband. "Wait," Barbara said, "why should my husband sign my consent form?" She would never forget the doctor's reply. "Because," he said, "women are too emotionally and irrationally tied to their breasts."[6]

Infuriating? Yes. But the exchange is also rather puzzling. Who was making this decision? Was it Barbara? We'd like to think so, but the doctor never actually asked her what she'd like to do. In fact, he said she was too emotional and irrational to be trusted with the choice. What does it mean to give someone the illusion of choice? If Barbara had said, "I don't want the biopsy," would the doctor have handed the consent form to her husband anyway?

"Looking back, I wish I had challenged that doctor," Barbara told me in 2015. "I wish I had torn up that office. I should have said, 'I wonder what part of the male anatomy men are irrationally attached to?'" She had many reasons to be angry, including the fact that the doctor doubted that women could make appropriate decisions under such stressful circumstances. But she didn't even question his presumption at the time. "I didn't think in those terms back then. Nobody did. That's what life was like for women."

This could just be a terrible moment in one woman's life, but as a decision-making researcher, I see larger concerns in this story. We would like to think that these sorts of things happened only in the past and that, at least in this very specific situation, the dynamic has improved. No doctor in the United States today would ask a woman's husband to make that decision. But how far have we really come? The temptation is to feel secure, to say such bias has disappeared. But how many of these biases about women as decision-makers have been fully erased and how many have merely gone underground, spoken of less often but still shaping who we want to lead? When making the wrong choice poses formidable risks, whether it's in the doctor's office or in a business meeting, are women seen as equals in the process or is there a creaking assumption that men are the ones with the superior decision-making powers, the gender that's unfettered by pesky emotions?

Almost half a century later, cancer treatment is a much more civilized process: women sign their own consent forms, surgery isn't scheduled until after the doctor and patient have discussed the biopsy results, and the radical mastectomy is largely a thing of the past. When Barbara tells this story now, everyone is appalled. But we have to ask ourselves, *Are things all that different?*

Richard Hoffman, a professor of medicine at the University of Iowa, finds that even today, there's cause to wonder what kinds of conversations take place between doctors and patients. What do doctors convey to their female patients and what do they suggest to the male ones about their roles in the decision-making process? Who gets asked, "What do you want to do?" Who doesn't? Are some patients treated as partners and others as dependents?

In 2011, Hoffman and his team analyzed survey data from eleven hundred adults across the United States, looking at patients' reports of recent conversations they'd had with their doctors about cancer screenings. Hoffman focused on adults over fifty because doctors normally recommend some types of regular cancer testing after that age. If physicians saw men and women as equally capable of making good choices, their decision-making conversations should have been the same regardless of the patient's gender. But they weren't. "Do you want to have this test?" was a question doctors reportedly asked 70 percent of men when discussing men's prostates but only 43 percent of the women when the visit was about women's breasts.[7]

Why this discrepancy? Why are men given more say in their testing options than women are when both are facing choices about cancer screening for sexual organs? There have been, it's worth noting, controversies around the effectiveness of prostate cancer screening. Initial screening for prostate cancer is usually done with a blood test, and approximately three out of four men who test positive don't actually have prostate cancer, meaning there's a high rate of false positives. Prostate cancer blood tests are considered so problematic that the U.S. Preventive Services Task Force gives the test a D rating, indicating it does more harm than good, whereas mammograms, which aren't perfect either, are at least given a B rating.[8] The prostate test can cause a lot of unnecessary worry, not to mention unnecessary procedures and risks from those procedures, which may be why doctors might give more men the option of whether they want to undergo this potentially misleading and upsetting screening process.[9]

Okay. So how about comparing apples to apples? Hoffman then looked at screening for an organ found in both sexes. He focused on a test that's received an A rating by the U.S. Preventive Services Task Force because it's such a reliable way to detect cancer in both men and women. The test? The dreaded colonoscopy.[10] Colon cancer is the third most common cause of death from cancer for both men and women in the United States, so it poses a high risk for both sexes.[11] When doctors talk with their patients about having a colonoscopy, do they simply say, "You need to do this," or do they present the options and then ask, "Do you want to do this?" Hoffman's results were revealing. Doctors asked 71 percent of the men whether they wanted a colonoscopy, but they asked only 57 percent of women. The numbers are better, true, but why aren't they identical? Why do more men get a choice? The men and women were in the same age range; most were between fifty and seventy years old, and the recommended age for the first colonoscopy is fifty. Men in the United States are at a slightly higher risk of developing colon cancer; one in twenty-one men faces it at some point in his life, compared to one in twenty-two for women.[12] But does that mean more men should have the option to decide for themselves, that men should be asked more often (rather than told) to take the test? A lower cancer risk for women suggests that if any gender should be given the option to skip the screening, it's women. And were female doctors more likely to ask women what they wanted (as opposed to telling them what to do)? We don't know. The data set didn't include information about the sex of the doctor.

When I first read this research report, I didn't know what to think. Maybe doctors were acting differently with men and women in some effort to be more effective. Physicians see hundreds of patients a year, and observing subtle patterns is part of any good professional's skill set. Could it be that physicians saw that men who weren't given a choice were offended and never came back? Or did doctors find that a large proportion of women avoided cancer screening initially and then later regretted it, so they gave fewer women a choice? Or is something less benevolent going on? Even though doctors no longer ask a man to approve his wife's surgery, they consistently seem to trust his good judgment (and their own) more than they trust hers.

The United States has a relatively short history of giving women the power to make decisions of consequence. Women in the United States weren't given the right to vote until 1920, after almost a dozen other countries had passed laws allowing women to participate in those civic decisions.[13] In 1968, when Barbara watched her husband reluctantly sign the consent form, doctors weren't the only ones who thought men had better judgment. Most of the professional world did. The women's liberation movement had just begun. Divorced women who tried to start their lives over in the late 1960s typically couldn't buy their own homes. A divorcée had two options: she could rent an apartment or, if she insisted on buying, she had to persuade a male in her life, often her ex-husband, to sign her mortgage.[14] When women with plenty of income applied for lines of credit in the early 1970s, they were often denied. Take Billie Jean King, the world champion tennis player who won three Wimbledon titles in a single year and supported her family on her winnings. She tried to get a credit card in her own name but couldn't. She discovered the only way she could secure a credit card was if her husband's name was listed first on the account; once it was clear to lenders that a man backed the financial decisions, she could be a secondary cardholder. If her husband had had an income, this might have made some sense, but he didn't. Billie Jean King was putting him through law school.[15]

Laws have changed lending practices, but why, in the twenty-first century, are these underlying sexist assumptions still affecting how women are treated with regard to decisions? Why might doctors involve men in decision-making processes more often than they involve women? Do most people, not just doctors, see men and women differently when sizing up a person's potential to make a good choice? Is there a little bias creeping into society's judgment of women's judgment?

To paint a better picture of what's happening, let's visit Victoria Brescoll's lab. Brescoll, a social psychologist at the Yale School of Management, is curious about how men and women are judged. She studies how ordinary adults appraise job candidates, potential students, and politicians. Brescoll wants to know how we determine whether the candidate on TV or the person standing in front of us is competent.

In one study, Brescoll asked adults to evaluate a job candidate. Each participant sat down in front of a video screen with a clipboard and watched a job interview already in progress. The interview seemed pretty routine until the manager asked the candidate (a male in one version, a female in another) to describe a mistake he or she had made at work—then things became interesting. The candidate told a story about working with a colleague and how the two of them had lost an important account. "And how did you feel?" the manager asked. "Well, I was pretty angry," the applicant replied in a loud, agitated voice, bristling visibly at the question. When the video finished, the subject had to rate the candidate. How much power did this person deserve in his or her future job? Would you trust this person to make independent decisions? Ultimately, should this person be hired?

Here's what Brescoll really wanted to know: Did the assessment vary depending on the gender of the job applicant?

It did. Brescoll and her colleague Eric Uhlmann, now a professor at INSEAD in Singapore, conducted three versions of this study, and every time, adults rated the female candidates who got angry lower than their male counterparts who did the same thing.[16] These weren't real job candidates, of course. These were professional actors, carefully following the same script, showing the same clenched jaw, venting at the same volume. But the people with the clipboards didn't see them that way. They saw the women who expressed anger as being "out of control," as individuals ill-equipped to take on leadership roles.[17] This wasn't just how male participants viewed women; it was also how women viewed women. Both sexes thought that women who expressed frustration about mistakes should have jobs with less authority and fewer opportunities to make independent decisions.

In contrast, observers chalked up the man's anger to stress, not to a lack of control. And expressing frustration about a mistake didn't hurt a man's credibility. Venting elevated his status. If an angry man mentioned that he was just a low-ranking assistant, then the people doing the evaluations wanted to give him more opportunities to lead. They wanted a frustrated man to make more, not fewer, decisions.[18]

Brescoll's research reveals that a man and a woman can say the same thing in the same way, and, without realizing our unconscious bias, we slip into thinking, *He's really onto something, but her, she's a handful.*

The Dogsled Problem

Some factions insist that women have plenty of decision-making authority now, and as evidence, they point to remarkable women in high places. Thankfully, women have made tremendous strides since the days when they were denied credit cards. In the United States, the last two judges appointed to the highest judicial positions in the country were both women, Supreme Court justices Sonia Sotomayor and Elena Kagan. Since 2000, citizens of Germany, Brazil, and South Korea have elected women to be their heads of state for the first time, making Angela Merkel, Dilma Rousseff, and Park Geun-hye not just three of the most powerful women in the world but three of the most powerful people. Even financial institutions that have been run by men since their inception have finally decided to give women a try. Janet Yellen began directing the central banking system in the United States in 2014, and Christine Lagarde was appointed to direct the International Monetary Fund in 2011.

Given all the names of powerful women people can rattle off, it sounds as though society now trusts a woman's judgment just as much as it trusts a man's. But these examples are more the exceptions than the rule. Women account for more than half of the world's population yet still make only a fraction of the big decisions when it comes to business and politics. Of the 195 independent countries in the world in 2015, only 11.2 percent of them were run by women. When we look at the five hundred largest companies in the United States and the one hundred largest in the United Kingdom, the percentage of female executives is sadly identical — only 15 percent.[19] Then there's what I call the John Statistic. When you look at the S&P 1500, there are more male CEOs named John than there are female CEOs of any name.[20]

Our perception of women as poorer decision-makers is not the only factor creating the gender gap in leadership. But just how much of a problem is it? Organizations like to give other reasons to explain the scarcity of women at the top, often pointing to an ambition gap, claiming that women aren't as interested as men in moving up the leadership ladder. But the data show that professional women are as ambitious as men. One 2013 survey of over fourteen hundred executives found that women wanted the top positions in the company just as much as men did. When the respondents were asked whether they wanted to be promoted to the next level of their organization, 83 percent of female executives said yes. Only 74 percent of male executives did.[21]

Other critics say that women aren't taking the right steps for success. They aren't finding mentors or sponsors, or they aren't asking for promotions or higher salaries, or they're not willing to sacrifice their evenings and weekends, whereas the men in the office can be expected to show up on a Sunday. Yet a study of over 3,345 graduates of the top business schools in the United States suggests "doing the right things" isn't enough for women. The nonprofit research organization Catalyst asked male and female business-school graduates which career strategies they used and then tracked, over several years, how much money these promising professionals earned and how quickly they advanced. Catalyst found that the supposedly tried-and-true career tactics helped men more than they helped women.[22] Women who went out of their way to adopt all of the best career-advancement practices still didn't move up as quickly or earn as much as men who used those same strategies.

Another camp shrugs and sees this as a pipeline issue. "Be patient," they say. "Now that more women occupy middle levels of leadership, in a few years, the system will catch up and more women will rise to the top." Maybe, but the statistics of the recent past don't support that.[23] The number of female board members in Fortune 500 companies hasn't seen a significant change for nine consecutive years.[24] The number of female state governors in 2014 was almost half what it was in 2004.[25] If women's progress is viewed as a pipeline, it's an extremely leaky one in science and technology; women leave the tech industry at twice the rate of men.[26] Kieran Snyder, the CEO of the software startup Textio, interviewed over seven hundred women who had left the technology industry and found that their primary reason for leaving was that the culture was unsupportive of women. One woman she interviewed, an electrical engineer named Jessica, said: "I loved my job. I loved solving problems and making technology that improved people's lives. What I did not love was going to work every day in a frat house."[27]

To understand the underrepresentation of women leaders, we need to ask ourselves, *Why wouldn't people trust women as top decision-makers?* It is not because women are seen as less intelligent. The Pew Research Center recently conducted a survey of over twenty-eight hundred American adults and found that 86 percent thought men and women were equally intelligent. Another 9 percent thought women had the intellectual upper hand. That means only 5 percent thought men were smarter.[28]

But *intelligent* isn't the same as *decisive*. What's our picture of the perfect decision-maker? Research shows most people believe leaders must be fearless and take action. They assume that those who don't flinch under pressure, who project confidence at every turn, make the best executive choices.[29] Bright people may know things, but good decision-makers bet heavily on what they know.

In most people's minds, all of the qualities above paint the picture of a man.

But when you look around your organization, you probably see plenty of women making important decisions. Thirty-eight percent of the managers in the United States are women. But what kinds of decisions do managers get to make?[30] Managers coordinate teams, plan projects, handle disputes, and oversee department budgets. For making these kinds of key internal decisions, women are often welcome. But when it comes to the choices that outwardly define the enterprise, that determine what an organization will be known for, we look to the head of the organization, most often a CEO. And for 95 percent of the largest companies in the United States, we're looking to a man.[31]

And some would argue that many of the decisions that middle

managers make aren't really their own. A 2015 study of 21,859 full-time employees found that supervisors and middle managers tend to be the most depressed employees, more depressed than both low-paid underlings and well-compensated business owners.[32] Why are people in the middle so unhappy? Because, the researchers at Columbia University and the University of Toronto reason, those middle managers have so little decision-making power. Supervisors and middle managers spend their time implementing other people's decisions; unlike workers on the front line, they can't easily blame someone else when things go wrong, and in many organizations, middle managers don't have the satisfaction of handling the final product or helping clients. Little autonomy, little evidence of impact, but plenty of pressure and responsibility.

When Liz, a visual artist and legal secretary in New York City, hears this, she's not surprised. She likens the division of decision-making throughout the professional world to dogsled racing. She believes women are welcome to implement all the supporting decisions in an organization and manage all the choices that assure the team will get to the starting line. People are happy to have women decide who needs more encouragement in the workplace, who needs to be disciplined, who might benefit from extra training, and who gets perks at the end of the day. They're fine with women managing the budget and even choosing who will be on the top team, explained Liz. "But on race day, when the spectators and cameras line up, it's a man who takes the reins, not the woman who orchestrated the entire thing." I call this the dogsled problem. Women might make the decisions that get the sleds ready to weather the course, but when the flag goes up, everyone expects men to be driving, making the visible, crucial decisions that win the race.

The history of men and women in computer programming pro-

vides a perfect example of the dogsled problem. As hard as it is to believe today, women were the early pioneers in computer programming.
Walter Isaacson captures the important and little-known role that
women played in the history of programming in his 2014 book *The Innovators*. In the mid-1940s, the engineers who built the hardware for
the first general-purpose computer were all men, but the programmers were all women. Grace Hopper, a female mathematician and na-
val officer working at Harvard, created COBOL, the first computer
program to use words instead of numbers. Men were more interested
in the circuitry and thought that programming amounted to secre-
tarial work, what with all that typing. There was no prestige; there were
no interesting decisions to be made. Programming was just preparing
things for race day, when the computer would solve a problem. Or so
it seemed to the men.[33]

But in the 1950s and 1960s, it became increasingly clear that there
were fascinating and important decisions to be made on the software
side. Engineers and administrators began to recognize that software
might actually be more important than hardware because a computer
program was transferable — it didn't run only on the computer where it
was developed, whether in Berkeley or Berlin; it could potentially run
on any piece of hardware. And suddenly, men wanted those programming jobs. Of course, the fact that programmers made pivotal choices
wasn't the only lure for men. There was a sharp increase in the proportion of men who majored in computer science in college around 1984,
and one hypothesis for this is that more teenage boys became interested in programming because parents were more likely to buy personal computers for their sons than for their daughters.[34]

Being a patient but ambitious hard worker isn't enough. Women will
keep running into the dogsled problem. If we want more women in
leadership, we need to stop seeing them as people who merely get the

team to the starting line; we need to start seeing them as potential drivers for race day and, even better, as leaders who can decide whether there should be a team in the first place.

He's Taking the Hard Road, but She's Making a Mistake

I'm a cognitive neuroscientist by training and for most of my intellectual life, I steered clear of gender issues. As a psychology major in college, I snapped up everything I could on language, memory, and reasoning and actively avoided classes on gender. If someone had asked why, I would have politely said I just wasn't interested in women's issues, but in truth, it seemed gendered to study gender. My male friends didn't research gender issues, and I wanted to show that I was just as mentally tough as they were. I shake my head at this thinking now, but when I was twenty-two, doing the most impressive thing mattered. In graduate school, I studied in three fields that were male-dominated — computational modeling, cognitive science, and cognitive neuroscience. I was usually outnumbered by men in the lab, often by at least six to one, but I never saw the scarcity of women as a problem. To suggest there was bias would have been signaling weakness. If I struggled, and I often did, it was my problem to solve — I simply needed to work harder. I was quiet and unassertive, but I didn't wonder if my scientific abilities were being scrutinized specifically because I had a ponytail. Scrutiny was the name of the game for everyone in the room.

I didn't think gender was a real issue, at least not in the professional world. So it took a baffling experience, all the more confusing because I'd doubted the problem even existed, to make me wonder whether a man's decisions might in fact be judged differently than a woman's.

When my husband and I became engaged, we lived three hundred miles apart. We both had promising jobs. He worked in Philadelphia and was being groomed for incredible professional opportunities, and

I worked in Pittsburgh in a role that was opening doors for me to take on increasingly high-profile projects. Some couples thrived in long-distance relationships, but not us. We liked to see each other at the end of the day, and this was long before Skype could make that even virtually possible. After a year of researching our options, looking for openings, discussing and re-discussing alternatives, we decided to follow my career because my boss had told me she was hoping I would one day replace her (and because the one comparable job offer I had in Philadelphia would have had me sharing an office with the photocopier, not the next CEO). Two days before our wedding, my soon-to-be-husband crammed his clothes, his computer, his trombone, and a Curious George monkey he'd had since childhood into his car, paid the tolls on the Pennsylvania Turnpike one last time, and moved into my apartment. We left for our honeymoon trusting he would find a job when we returned.

He did find employment, but it was a terrible, demoralizing fit. He worked hard and committed to making the most of the situation — we even put a bid on a cute little Colonial house — but after about a year of trying, we could see that he needed a different avenue. We faced a dreadful decision. Over the course of that year, I'd enjoyed a lot of recognition at work. I had received a sizable raise and coauthored two grant proposals, and my boss had invited me to lead a rewarding research project. But after more discussions with my husband and more investigation, I had to admit that, as sad as I was to leave my work, it was only fair that we give his career the same due we had given mine. We picked a part of the country where I wanted to live, he applied for his dream job, and the company made a good offer. I didn't have a job yet, but just as we'd done for him, we trusted that I'd find one.

A year earlier, when my husband announced to his colleagues that he was quitting his job to follow me — that he was leaving a great employer to go somewhere without a prospect of a new job — several acquaintances at work told him he was making a noble decision. His supervisors said that he was always welcome back, and if he needed a recommendation, they would gladly put in a good word. None of his friends, family, or coworkers questioned his decision, at least not to his face. No one said, *Are you sure you want to do this?*

What happened when our situations changed and I decided to join him as he followed his next big career opportunity? I wish I could say I met with the same reception he had. Instead, my boss told me she didn't think I was taking my career seriously enough: "Too many women do this. Don't make this mistake." A few acquaintances and several good friends asked, with a great deal of concern, if I was sure I was making the "right choice." My boss wrote a glowing recommendation letter praising the work I'd done, but she came into my office one day, closed the door, and said she no longer thought I was ready for the management position she'd been grooming me for, not if I had this kind of judgment.

Did I chalk this up to gender bias? Not at all. Instead, I assumed it revealed a deep and unsettling truth about me. Maybe those concerned friends saw something disappointing; maybe this choice indicated that I wasn't that serious about my career after all. Maybe this was a sign that when push came to shove, I'd be traditional and put my husband's needs first. I'd thought we were taking turns: my career, then his. But maybe this was how couples always started the negotiation before they slipped into the typical pattern: his career, then his career again. What bothered me most was my supervisor's comments. Perhaps she was right—perhaps this did reveal a major flaw in my decision-making abilities and I wasn't ready to be a manager. I spent many late nights wondering what perfect line I might have walked to hold on to her respect and still support my husband. I also began to question my own judgment. When I did find a great job—as a manager, actually—I downplayed my new status. The nameplate on my door said DIREC-

TOR, but I avoided calling myself that, describing my position in ambiguous ways, such as "I work at the teaching center." Part of me feared that if I mentioned my title, people would question whether I deserved to lead.

My husband was the one to observe, months later, that perhaps gender had played a role in our varying experiences. "Isn't that interesting?" he said. "A man decides to set aside his career to follow his wife and no one doubts him, but a woman makes the same decision to follow her husband and she's subject to scrutiny." I stopped in my tracks. Sure enough, when I made my decision, I had felt pegged as someone who was letting women — and certainly myself — down. With that single choice, I'd brought my entire history of strong decision-making into question. But when my husband made his choice, he was patted on the back for taking the hard and admirable road, for supporting his spouse. Was this unique to the two of us, something about the way my husband and I described our decisions that made people doubt mine and salute his? Or was this the tip of something bigger?

It can be hard to accept that decision-making is stickier for women than for men. We want to believe in a just world. We want to believe in progress. My friends and the boss who had misgivings about my decision probably had the best of intentions. They might have been looking out for me, but are we quicker to judge women's choices? Society's beliefs about men's and women's decisions often lead people to put women's decisions into constrictive, stereotyped boxes. That might be comfortable for society but it's deeply uncomfortable for women.

Why Women, in Particular, Need to Understand Decision-Making

Great decision-making isn't a single skill — it involves many skills. You have to be able to clarify your priorities, come up with options, analyze those options, test your assumptions, select the option that matches your top priorities, generate buy-in, and prepare for what happens if you're wrong.[35] Both men and women can acquire all of these skills, but as we'll see in this book, a few of them come more easily to women.

It's not difficult to see why organizations value resourceful decision-makers, men and women alike. In an increasingly 24/7 work culture, bosses want employees who exhibit good judgment, who don't have to check with their supervisors at every turn. But as we'll see in this book, having good decision-making skills is especially important for women in the workplace because women typically have to prove themselves more often than men. If a man makes one pivotal good decision for his company, it can carry him for a long time. He might be promoted on the basis of that one crucial call. But most women need to show that making a smart and strategic choice is repeatable, that it wasn't just luck or timing or connections that led them down the right path. Women face what Joan Williams and Rachel Dempsey call the "prove it again" bias.[36] Stellar judgment has to be a skill a woman possesses and can use again.

Many women are also fighting the perception that they're less likely than men to commit to a course of action. There's a popular misconception that women are indecisive by nature, that, unlike men, women continuously review their options, avoiding the responsibility of choice. But it's not a fair assumption. As we'll see in the chapter on decisiveness, a woman's colleagues might mistake a desire to collaborate for an inability to decide.

A Different Playbook

Brilliant books have been published on decision-making, but some of their advice is great for men and terrible for women. Consider these recommendations: "Whenever possible, get everyone to agree on a decision" and "Take more time to consider a fuller array of options." These strategies might seem refreshing when adopted by a male executive, but not by a female one. If a woman vice president says, "Let's not decide until everyone is on board," people will assume that she's insecure and incompetent, not that she's attempting to build a coalition. When a woman takes her time to analyze options, she's likely to be called indecisive and underqualified rather than considered and thoughtful.

Women, even the most highly respected, face a different set of expectations and stereotypes than men, so they need decision-making advice tailored to match. It's time we stopped telling women to use a man's playbook.[37]

Perhaps even more troubling? Many of the tips in leadership books that are designed for and target women don't incorporate what scientists have learned about how great decisions are made. "Be as confident as the bold men around you" is a common piece of advice for women, but studies show confidence actually stands in the way of good judgment and is a root cause of many poor decisions — for men *and* women. Being more confident might help people speak up in meetings, but it doesn't improve the quality of their contributions. Confidence is a tool all of us can use, but the most adept decision-makers know when to pick it up and when to put it down.

What you won't find in any of these books for men or women are the many discoveries researchers have made about gender and decision-making that could be put to practical use in organizations. A growing body of evidence shows that when men and women make decisions under intensely stressful conditions, women's choices tend to veer in one specific direction while men's go the opposite way. These counterbalancing reactions mean that to make good decisions under stressful conditions, we need both men and women in the room, and both need to be heard. Neuroscientists know about these biases, but investment bankers probably don't. Public-school superintendents probably don't. The more people learn about the science of gender and decision-making, the better prepared and more mindful they can be when they set up teams that have to work under pressure.

This book is meant for any woman who cares about her career. Women at all stages in their professional journeys — whether they're applying for their first full-time jobs or considering a leave of absence midcareer — should find answers here. I've chosen to focus on women's decisions in the workplace because there's a disconnect in the literature and because so many women suspect that they're to blame for the way their decisions are judged once they walk into the office; they wonder if their career challenges and sense of disadvantage are of their own making or if there's something systemic going on and, if there is, if anything can be done to destabilize the current dynamic. But I want all women to feel included in the conversation I hope to start, and the questions and strategies offered here apply to life beyond the workplace.

This book is also meant for men who want to make their organizations as strong as possible and who care about the women around them. If you're a man who wishes you could hold on to the talented women in your organization, you'll learn what they're up against and find strategies for changing the environment so their decisions are judged fairly and they have the best chance for succeeding. And if you have a daughter and you want to understand why her decision-making process looks rather different from yours, this book is for you.

I've also written this book for anyone who wants to make time at work more productive. As professionals, we go to seminars on increasing our productivity, and we go to seminars on harnessing diversity, but rarely do we see a connection between the two. What if you see that connection? What if being inclusive increases the quality of your decisions while still allowing you to meet your deadlines? There is no magic wand included with the purchase of this book, no recipe for a secret potion on the inside of the back cover (or anywhere, that I know of). But there are concrete suggestions for what you can do to make sure more good ideas surface and ensure that your colleagues — women and men — are able to bring their best to the decision-making process.

The Decision Dialogues

Looking back on how strangely I had been treated in my decision-making paradox, I wanted to know how women's decisions worked across the board. Was it common for decisions to be judged by a different standard depending on the decider's gender? Were men and women at all different in how they arrived at their choices? I'll admit, I didn't want to believe in the popular caricatures, the image of men making swift decisions and women taking forever to make up their minds, but I had to ask — were such notions true?

I started this project by gathering small groups of women I respected and asking them to talk about decisions they were pleased with and decisions that gave them pause. Decision Dialogues, I called them. I asked questions. I took notes. And I fell into quiet awe at what I heard. The thoughts and feelings these women had about their decisions were far more complex than I would have gleaned from any cocktail-party conversation or academic study. One woman talked about how difficult it was to decide whether to become the power of attorney for her dementia-suffering father. Another woman talked about quitting her career in nursing to go back to graduate school. Several women of Asian heritage said they envied the freedom the others in the room took for granted — even though the Asian women were accomplished adults, their parents had unreasonable expectations they had to manage. Whether their brothers had to shoulder identical burdens, I don't know, but the women found the expectations exhausting.

The conversation always circled back to how women's decisions were seen by others. Some women said they didn't care what others thought; others admitted that they cared deeply how they were perceived, and many were torn. But whatever the participant's stance, the topic of how decisions were judged ignited discussion.

I scoured the popular books on decision-making. These books didn't talk about gender or about making decisions under scrutiny or about how to manage that scrutiny. In addition to reading the entire pop-culture decision-making library, I combed through hundreds of primary studies on how women and men made choices and the complex space they both entered every time they faced a decision with serious stakes. Then I reached out to and interviewed women directly. Thirty-four women offered their own experiences and helped build the stories that illustrate the lessons and insights in this book. I've used pseudonyms for all of them except Barbara Winslow and a few others who are public figures and who have shared parts of their stories before. You'll meet a police chief, a head chef, and a punk rocker. You'll hear from a mathematical genius who specializes in neural theory and a woman who started a company because she was fed up with the typical dating websites. You'll watch a woman going for an endurance world record decide if she should keep pushing even though she can barely move. Some of the women who spoke to me were in their twenties and staring down their first decisions after college, others were in the prime of their careers, and a few were retired and looking back at the choices they'd made.

As you read about some of the findings in this book, you're bound to think of exceptions. You may be the exception. Or your wife may be. Perhaps your best friend from childhood grew up to be a fighter pilot, and she personally defies the entire chapter on gender and risk-taking. That's okay. We can't say that all women decide this way and all men decide that way any more than we can say that all women have longer hair than all men. As you read the research that follows, if you keep the phrase *on average* in mind, you'll get more out of it. It's rarely the case that every single member of one group differs in exactly the same way from every single member of another group on anything, whether the issue is how often someone interrupts or how often a person makes a poor decision.

Put Away the Duct Tape—This Isn't About Fixing Women

The first task of this book is to examine how we see women as decision-makers. What do people, women and men, believe is different? What do we see as the strengths of women's decision-making and what do we see as the weaknesses? All too often, women look back on their own decisions, choices that were complex and thoughtful, and shoehorn them into a narrow wedge that makes the process seem simple and obvious.

Some people are quick to say—perhaps not publicly, but privately in their offices and anonymously on websites—that there's a grain of truth behind every stereotype. These individuals insist that when they believe something negative about women, it's because they have eyes and ears and know what they've seen and heard. "I wish it weren't true," they cluck, "but it's certainly been my experience." And that's the question that makes us all hold our breath when we discuss group differences. Is there a grain of truth in the stereotype, or are we noticing only what we expect to and forgetting all of the exceptions we've encountered? What does the science say? Examining that question, pinpointing and understanding the science, is the second task of this book. To do that, we'll investigate research findings and stories of real women to explore ways that women are misunderstood as decision-makers. We'll discover when men and women follow the same steps in making decisions and when they usually take different routes, and we'll learn what motivates them to part ways.

I want to be clear, though, that there are two things this book is not. First, it is not a book about fixing women. There have been several wonderful books about women in the workplace in recent years, but a recurring theme in the conversations they've sparked is "women are doing the wrong things." When I look at how women make decisions, *wrong* is not the word that comes to mind. *Brave?* Yes. *Flawed?* Sure — but no more than men. *Misunderstood?* Absolutely. This book is meant to help us all understand how women make decisions, how they navigate judgments in professional environments, and how this understanding can be used to make the most of their skills and talents.

Second, this book won't tell women which choice is best. If you're a woman who's trying to decide whether to stay in your career or go back to school, you won't find definitive answers here. I can't tell you whether to marry the person you're dating and I certainly don't know if being an artist is right for you. This isn't a book about *what* women decide; it is a book about *how* women decide.

Still, I will offer practical strategies for making well-informed choices, and I'll offer advice that both men and women can use. If you're under a lot of pressure to make a decision, is there a way to use that pressure to actually improve your thinking? Is there ever a time when it's wise to just go with your gut? By offering strategies that lead to better decisions, I hope readers of both genders will find their way to fulfilling — and well-respected — choices.

It's a good time to reexamine how we think about women and decision-making. We're coming up on the year 2020, which marks the one hundredth anniversary of women getting the right to vote in the

United States. Even though it has been almost a century since American women were first afforded a chance to say "I want this person to lead, not that one," society as a whole still has surprisingly antiquated notions about how women decide. The gap between the modern world and our outdated beliefs inspires admiration for women who have made bold decisions and withstood harsh judgment, and kindles compassion for those who have repeatedly second-guessed their own ability to make strong, informed choices.

Expanding a Woman's Toolkit

Remember Barbara Winslow? She woke up from the biopsy to the best kind of news. The lump was benign. She hadn't undergone a radical mastectomy in her sleep. She also woke up to realize that she didn't want to be the kind of woman who went through life undervaluing her own judgment. The next time a doctor disrespected her, only a few years later, Barbara stood up, grabbed her coat, and left. She went on to become an international expert on women's activism, and in 2012, the National Council for Research on Women named her one of its Thirty Women Who Make a Difference.

How do women expand their decision-making toolkits? How can women stop second-guessing their judgments and start second-guessing what they're hearing about women's ability to decide? They don't have to follow directly in Barbara's footsteps and become celebrated activists. But they can take a good look at what they've been assuming about their thought processes. They can choose the decision-making tools they want to pick up and recognize and set aside the decision-making biases they've held on to for far too long. This book is an invitation and a toolkit for women who want to make stronger, smarter decisions in a world that still whispers that they can't.

1

Making Sense of Women's Intuition

WHEN ISABELLE WALKED IN, she struck me as stylish and slightly shy. I expected her to be wearing a starched chef's jacket and have the constantly moving energy of a cooking-show personality. Instead, she was dressed in a smooth, tan silk suit and stood with clasped hands as she quietly asked me if I had any trouble finding the place. We stepped into a conference room at the oldest and most famous culinary-arts school in the world, Le Cordon Bleu Paris, and she offered me a pot of perfectly brewed tea. I privately hoped that there might be a plate of profiteroles somewhere, but I wasn't there to eat cake. I was there because Isabelle was one of the most powerful women in the French culinary world.

As one of the Cordon Bleu's top administrators, Isabelle decides who will be hired to instruct the next generation of chefs. She chooses which recipes students will have to master before they're certified to cook for the world with the stamp of approval from the Cordon Bleu, and she selects the chefs who will participate in Europe's star-studded food competitions. She's called a resource center manager, but she's really a decision manager.

I asked Isabelle to tell me how she decided to come to the Cordon Bleu. She had been the head chef at an embassy in Europe for a decade, and over the course of her career, she'd prepared meals for everyone from Julia Child to France's prime minister. Not a bad job by any measure. She explained: "I got a call from the president of the Cordon Bleu at the time, and he said, 'Isabelle, I want you to come work for me. I want you to manage the Cordon Bleu.'" She shrugged and smiled at me. "Everyone loves hearing that." Then she sighed and stared at a spot on the floor. "But it still wasn't an easy decision. I knew I'd be gaining a lot, true, and I wouldn't have to be on my feet until two thirty in the morning preparing a cocktail party for five hundred. I was getting too old for that kind of physical labor. A desk job was looking pretty good," she said. "But I'd also be losing some friends, all of my connections, some real perks. I wouldn't be cooking every day and I'd miss that. I talked with several people who had worked here at Le Cordon Bleu Paris to get a real sense of the place. I spent two, no, really three, months thinking about it, and I finally decided yes, I can do that job."

I was surprised and impressed. Three months is longer than most of us can sit with a job offer, no matter how much we're wanted. But what really shocked me was what Isabelle said next. She looked up from the floor, nodded, and said proudly and without a hint of irony: "You could say I just went with my instincts."

Just went with her instincts? Months of deliberation, of weighing all the advantages and disadvantages of a job, of researching the reality of it, is hardly a gut, instinctive reaction. I heard this pattern again and again in my interviews with women. In relationship decisions, in job decisions, in medical decisions, women think long and hard about their choices, but once they've made those choices, even well-educated, powerful, strong women put them in a box labeled *Going with My Gut; Following My Heart, Not My Head;* or *Trusting My Intuition.*

Are all of these women just being humble? What I found most revealing is that these were the decisions the women were most pleased with. When they looked back on their lives, those decisions that they had analyzed so carefully, that they had turned over in their minds again and again and yet ultimately attributed to instincts or intuition, ranked as some of the best choices they'd ever made.

Have you ever heard the term *men's intuition*? Even once? Probably not. Intuition has been branded a woman's thing, as though it's something worn under a bra. Popular news blogs such as the *Huffington Post* routinely run articles about the strength of women's intuition, as do publications that target women, such as *O, the Oprah Magazine*. In 2015, a columnist for *Psychology Today* advised, "Women, practice this mantra: trust your inner knowledge, your intuition, that gut feeling . . . If you don't trust someone when making a deal, go with that feeling." But admit it? No. The article warns that women shouldn't reveal that they're making decisions based on their "intuitive radar." Remember, it counseled, "feelings are not credible in a man's world."[1]

One reaction I have when I read columns like this is that there's a striking disconnect — women are told to be proud of going with their gut, but they should keep it under wraps. Own it, but own it hush-hush. Is this supposed to be a secret superpower? But that brings me to other questions. Do women actually rely on intuition more often than men? And should they? Is women's intuition a legitimate phenomenon? And when should anyone — man or woman — rely on gut feelings alone to make a decision?

We're going to inspect a notion that many people hold dear, an ability that some women believe sets them apart in a good way. A few of them might read this chapter and think, *Can't you leave the upsides of being female alone and let us win one?* But as you'll see, the concept of women's intuition isn't as empowering as it first appears.

Do You Go with Your Gut or Your Head?

Before we dig any further, it's helpful for you to assess your own cognitive style and see where you fall on the intuitive-analytical spectrum. When you face choices in your everyday life, are you more inclined to be intuitive and go with your gut or do you prefer to be analytical and go with your head? I've developed an informal questionnaire to help you identify which approach you favor. First, make a numbered list from 1 to 14. (It's okay; you can write in the margins.) Answer each question True or False. You'll answer True if the sentence describes an approach that comes naturally to you or the approach you prefer when it's available. You'll answer False if the sentence sounds like a poor fit for you, if it describes something that doesn't come naturally to you, or if it's the approach you'd take less often if you could. Think back to decisions you've made in the past three months, everything from whether to buy a fitness tracker to whether to take on a particular client. Rest assured, there are no right or wrong answers. You'll probably have stronger responses to some questions than others; for those you feel lukewarm about, do your best to give it a True or a False.

1. I am more at home with detailed facts and figures than with broad descriptions and ideas.
2. My gut feeling is just as good a basis for decision-making as careful analysis.
3. I find that adopting a multistep, analytical approach to making a decision often takes too long.
4. I am inclined to scan through reports rather than read them in detail.
5. I work best with people who are spontaneous.
6. I rarely make off-the-top-of-my-head decisions.

7. Projects that require a logical, step-by-step approach feel constraining to me.
8. I prefer to solve problems by gathering data.
9. I often wait to make a suggestion until I have some evidence to back it up.
10. I find that it's possible to be too organized when performing certain kinds of tasks.
11. People have sometimes told me, "You might be overthinking this."
12. I am most effective when there's a clear sequence of tasks to be performed.
13. My philosophy is that it is better to be safe than risk being sorry.
14. I think the best problem-solving happens by bouncing between different ideas and possibilities.

Now take a moment to calculate your score. Use the following key: For items 2, 3, 4, 5, 7, 10, and 14, give yourself one point for every question that you answered True. For items 1, 6, 8, 9, 11, 12, and 13, give yourself one point for every question that you answered False.* Now add up your points.[2]

Your score might be as low as 0 or as high as 14, but don't worry if you scored a 0 or a 1. A low score here is not a bad thing. A low score means you are probably more analytic in your approach to problem-solving and decision-making, and a high score means that you're probably more intuitive in your cognitive style. If you scored right in the middle, 6, 7, or 8, researchers would call you adaptive, which means

---

* Keep in mind, this is an informal quiz, not a scientifically validated test of personality. I've formulated the questions by adapting items from two commonly used assessments of cognitive style. (For a longer discussion of different measures of intuitive styles, see the endnotes.)

you blend analytic and intuitive styles and use elements of each depending on the circumstances.

But whether you take a highly intuitive approach or you're happiest being analytical, when you do find yourself trusting intuition, what exactly are you depending on?

An Inner Compass or a Firefighter's Well-Trained Senses?

How do most of us think about intuition? An intuition feels like something you receive, something that appears immediately and perhaps unbidden, not something you worked all day to generate. Brilliant minds have credited intuition for their great discoveries. Jonas Salk, who discovered the first successful polio vaccine, said, "It is always with excitement that I wake up in the morning wondering what my intuition will toss up to me, like gifts from the sea. I work with it and rely on it. It's my partner."[3] Oprah Winfrey, one of America's favorite talk-show hosts, has referred to intuition as the "inner GPS guiding you to the true North" and the "small voice" that you'll hear only if you sit still and just listen.[4] For many of us, an intuition is a sudden feeling that you know something that goes beyond the data in front of you, a tugging conviction that, in the middle of a decision, you should just ignore the outside world for a while and tune in to that voice.

I keep calling intuition a feeling, but that word doesn't do it justice. Intuition isn't like envy or excitement or some other run-of-the-mill human emotion. Intuition feels like knowing. Best of all, it feels like knowing without any effort. Robert Graves, the English poet and novelist, described intuition as "the supra-logic that cuts out all the routine processes of thought and leaps straight from the problem to the answer."[5] If we break it down, the word *intuition* means "being taught or guided from the inside," and because it comes from inside, not out there, it feels more accurate, more deserving of your deepest trust.[6] We feel drawn to listen to it, to follow it, and if we experience a gut instinct but choose to ignore it and turn the other way, we wonder if we've gone against our better nature. If somewhere down the road it's clear that, yes, we did choose poorly, we confide to our closest friends that we "knew" the right choice from the beginning. We just didn't listen.

This is how most of us think about intuition, and the scientists who poke and prod intuitions on a daily basis find that our guesses are right about four key things. First, they agree intuitions are fast. Designers can look at a handbag and judge if it's the real deal or a knockoff Prada in less than two seconds.[7] Expert firefighters can walk into a burning building and, within minutes, intuit the part of the building that's about to collapse.[8] Scientists also agree that intuitions are usually accompanied by strong emotions, a feeling of being drawn to one thing or repulsed by something else. One fire-team commander in Cleveland, Ohio, was hosing down a kitchen fire with his crew when a bad feeling suddenly came over him. He didn't know why, but he found himself shouting, "Let's get out of here." The kitchen floor caved in seconds later. Investigators learned it wasn't just the kitchen that was on fire but the entire basement below.[9] Most of us aren't firefighters, but we all know what an inexplicable repulsion or attraction feels like.

Researchers also agree on a third defining quality of intuitions: they involve making holistic connections, surprisingly large leaps between something small that's in front of you right now and something you already know well.

In the case of the fire-team commander, the fact that he was covered head to toe in safety gear but had raised the earflaps on his helmet helped him make that leap. The researcher Gary Klein interviewed him multiple times after the fire, and with some careful sleuthing, they pinpointed two things the commander probably reacted to: first, the firefighter noticed his ears were very hot, which he wouldn't have expected for such a moderately sized fire, and second, the fire was quiet given the amount of heat it was generating. Most of us wouldn't notice those two small cues—who knows how loud a kitchen fire should be?—but the commander's years of experience told him something was very wrong.

Learning to squeeze the most from scraps of information is something that Amanda, a surgeon at Harborview Medical Center in Seattle, Washington, trains her medical residents to do. Amanda works in a level-one trauma center, the only one of its kind across several states. A level-one trauma center receives the sickest and most severely wounded patients. Ambulances rush one-year-olds with third-degree burns to Harborview, and helicopters bring construction workers with crushed pelvic bones to Amanda's care. Patients are flown in from Montana and Alaska, over a thousand miles away. Several times a week, she and her team need to make rapid decisions about whether there's time to get a CT scan or whether the patient needs to be rushed to surgery without any further information. Skilled observations and expert intuitions are crucial. Amanda trains her residents to make these fast, accurate judgments with minimal information in various ways. She tells them that every time they walk into an ICU patient's room, they need to stand for a moment in the doorway and, without looking at the chart, take in all they can and guess whether the person lying in the bed is doing better than he or she was yesterday. "It's just a guess," she explained, "and in a moment, they'll get to look at the patient's chart and learn whether their guess was right or wrong." Amanda wants them to tune in to the signs, to notice the subtle cues they can connect with the vast textbook knowledge they already have. She needs her fellow physicians to recognize when they can see a little but know a lot.[10] That's the definition of a good intuition.

What Amanda is doing is innovative because most medical training involves learning to articulate exactly what you see, what you don't see, and what that pattern of symptoms suggests. But Amanda isn't trying to train physicians' conscious thinking — those doctors wouldn't be at her hospital if they didn't have a great deal of intellectual ability. She's training their unconscious, helping them practice snap judgments. And most experts agree that intuition doesn't arise from conscious, deliberate, or rational thought. We don't plan the steps of an intuition, nor can we typically retrace them, so we often don't know where an intuition or gut reaction comes from.

Sometimes it's hard to accept that we can't consciously explain our gut reactions. We can always generate a reason. Let's say you don't like the woman who lives two houses down. If you walk out the door and notice her, you'll step back inside. You see her only once a week and you can't quite put your finger on why you don't trust her. If someone gave you a minute, you could probably come up with a socially acceptable reason, even if it wasn't what actually triggered your dislike. She never says hello, you'll say, even though you can't think of a time when you said hello either. Or she lets her trash bin overflow. But all of this is in hindsight. You're looking back and trying to justify why you felt or behaved a certain way, but if you're honest, you really don't know. You aren't repelled by consciously reviewing the data. You don't watch all the little levers falling into place, like a complex Rube Goldberg machine. You just see the end result, the ball that rolls out into the little chute of consciousness, and you feel disdain.

As a general rule, our everyday understanding of intuition lines up well with the science. But according to researchers, most people are wrong about one thing when they talk about intuition, and it's perhaps the single most important thing. All of us assume that our hunches will naturally lead us to better choices and that our gut feelings sniff out essential elements that our conscious minds miss. Sometimes that's true, and sometimes it's not. I wish I had better news, but gut instincts aren't very astute. To some extent, you probably already know this. At least, you know that *other* people's intuitions are often wrong. Your uncle puts several thousand dollars in an investment opportunity because it feels right, and when you ask if he did any research on the company, he shrugs and changes the topic. If it was your mom who couldn't stand the woman two doors down, you'd say, "I don't see anything wrong with her — maybe it's all in your head." As we'll see in this chapter, research over the past two decades reveals which kinds of intuitions are the most trustworthy and which deserve more systematic scrutiny.

Women Lean into the Data

So is women's intuition everything it's cracked up to be? This is a good question, but there are two better questions hidden beneath it. First, are women more likely than men to use intuition when they face a choice? The stereotype suggests that women make decisions intuitively, based on their feelings or something they can't explain, while men make decisions analytically, based on a linear thinking process that's backed by data or a logical argument they can show you in a PowerPoint presentation. We deserve to know: Is there any truth behind this popular notion?

The short answer? No. Christopher Allinson and John Hayes, management professors at the University of Leeds, analyzed thirty-two research studies that compared the decision-making styles of men and women. These studies mainly focused on people in business, but they included people in different locations and at different stages in their careers, from undergraduate business students in the United States to upper-level managers in Singapore. All of these studies used the same diagnostic test of decision-making style, the Cognitive Style Inventory, which was developed by Allinson and Hayes to evaluate intuitive and analytic decision-making styles. (The questionnaire you took earlier in this chapter was adapted, in part, from their inventory.) And their analysis found that women weren't necessarily intuitive. Forty percent of the studies concluded that women adopted a more analytical style than men, the exact opposite of what many people think. What about the remaining 60 percent? The rest of the studies found no significant difference between men's and women's cognitive styles, showing that men and women were equally intuitive (or analytical, as the case may be) in their thinking process. None of the thirty-two studies, not a single one, found that women tended to have a more intuitive decision-making style than men.[11] In almost six out of ten studies, there was no difference in who went with their gut-level reactions and who went with their heads.

You might not be surprised to hear that if anyone's more analytical and systematic in their thinking in the workplace, it's women, not men. Many women feel they need to have all of their ducks in a row long before they propose an idea, particularly if they're working with men and want to be sure they're heard. Kat is a former CEO of a technology company where, like most technology companies, the men outnumbered the women. She found women tended to point to the market research when they wanted to pursue a new feature for one of the company's products or take the business in a different direction. But the men? Kat said, "The men liked to think that they had just come up with an idea." The men saw themselves as visionary, she said, while the women saw themselves as justified.

Studies show that if men and women are having a discussion and there's no designated leader, the men enjoy higher status in the conversation unless the topic is clothes-shopping or child-rearing or something that's traditionally accepted as an area of feminine expertise.[12] Consider what happens when a man and a woman are working together on a contentious task, such as deciding where to make budget cuts. When the two disagree on how to solve a problem, sociologists at Bowling Green State University have found that even if the man made noticeable mistakes earlier in the conversation and it's evident that the woman knows more about this topic than he does, the woman will still listen to what the man has to say and allow him to influence her judgment. But the same isn't true if the woman made visible mistakes. According to the study, once a woman is found to be wrong in a disagreement, the man is more likely to stick with his view. Her misstep is more costly than his. These findings suggest that women are more successful in persuading and influencing men only if they've previously outperformed them.[13] So when women think through every angle of a decision, it could be because they want to be sure they're taken seriously.

There's also the factor of what a job requires. Some scientists argue that if you want to predict whether a person is going to be more intuitive or more analytical, look at the job, not the gender.[14] If you're a stock trader on Wall Street, you're going to be asked to make fast-paced, intuitive decisions regularly because on a volatile day, a stock price might change three times before you can finish reading the morning headlines.[15] But if you're an actuary working for an insurance company deciding whether to raise or lower insurance premiums for all homeowners under thirty, gut feelings won't cut it. In that kind of job, your boss will expect you to back up your recommendations with detailed analyses.

Want to Make a Team Smarter?
Add Women and Genuinely Include Them

So, contrary to popular belief, women tend to base their choices on documented analyses at least as often, if not more often, than men. But some people mean something very different when they say "women's intuition." Some believe that women are better at reading people, and they use the term *women's intuition* to refer to an increased sensitivity to verbal and nonverbal cues that help one person infer another's emotions.[16] A sensitive person walks into a meeting, scans the people who are already seated, and, before anyone says a word, thinks, *Steve is really angry about something. This meeting is going to be harder than I thought.* Discussion begins, and sure enough, Steve is exasperated. Here, *women's intuition* refers to empathy, what researchers refer to as interpersonal sensitivity, social sensitivity, or empathic accuracy. So that's the second question — are women typically better than men at reading other people's feelings and therefore more intuitive in their judgments?

If you ask your work colleagues how well they read other people, you'll probably find that women are more likely than men to say they're pretty good at it.[17] Society tells women they're supposed to be mind readers, antennae quivering to decipher the thoughts and feelings of others.[18] On the classic TV sitcom *Friends,* Chandler and Monica often had conversations that looked like a game of charades, except that Monica did all the guessing. Chandler would look upset, and Monica would guess that something awful had happened. Chandler would make a little hand gesture that meant "keep going," and she would. She'd offer more and more complex scenarios and Chandler would just nod until Monica had accurately determined an entire sequence of events based on nothing but his eye-rolling.

Was Chandler expected to extrapolate from Monica's sour looks? No. And when he did, he usually got it wrong. It's just a television sitcom, but it shows how often we're all told and how easily we accept that men should have a free pass when it comes to interpreting what other people think and feel. Women's magazines often advise their readers not to give a male coworker, boyfriend, or husband a hard time if he has trouble reading emotions. In one bestseller about gender differences, the author maintains that women often know men's emotions before men themselves do.[19] The message we hear again and again is that women have the empathic advantage.

And this is one message that's partly true, at least the part about women's skill in deciphering other people's emotions. The claim that women intuit men's emotions before men have a clue about them exaggerates women's power of perception and insults men's self-awareness. (Besides, as we'll see soon, men aren't doing themselves any favors by taking a free pass in this arena.) Research indicates that most women are more adept than men at interpreting nonverbal cues, so women might be quicker to notice that a colleague is feeling impatient with a conversation and more accurate in determining whether the boss's slumped posture means defeat or concentration. Women are also quicker to gauge by the sound of someone's voice if that person is angry, which can be a helpful piece of information if someone is trying to make a decision in a conference call and can't see anyone's face.[20]

Perhaps the most peculiar and bizarre finding in this area of research is how women perform on the Reading the Mind in the Eyes Test, or, more simply, the Eyes Test. Anita Williams Woolley, an organizational psychologist at Carnegie Mellon University, and her colleagues asked men and women to guess the emotions of a person when all they could see was a narrow strip of a face, a horizontal window that showed the eyebrows, the eyes, and the very top of the cheeks.[21] Are those eyes worried or annoyed? Amused or sarcastic? It sounds impossible, and few people do it perfectly, but Woolley and her colleagues found that women outperformed men across multiple studies.[22] It's not very often, of course, that anyone has to judge another person's mood on this isolated piece of information — usually if you can see someone's eyes, you can also see if that person is grimacing or smiling or smirking.

So are women "naturally gifted" at reading emotions? No — evidence shows that men can read people as well as women do, and it doesn't require attending a Saturday seminar on interpreting body language. Expectations and motivation are key. Researchers have found that men can read emotions as well as the women around them when they believe their cognitive ability, rather than their emotional sensitivity, is being tested.[23] If it's about intelligence, he's all in. Tell people that empathy is being tested, however, and women's scores rise above men's. Likewise, when college-age men believed that the ability to read another person's feelings would lead to more sex, their accuracy at reading emotions improved immediately, no training necessary.[24] So with the proper incentives, men's abilities rival women's, suggesting that motivation is clearly a key part of the story.[25] And it's not just men's incentives. Sara Hodges, Sean Laurent, and Karyn Lewis, a team of psychologists at the University of Oregon, believe that one of the main reasons women are so skilled in tests of empathy is that they have more on the line.[26] Put a woman in a test of interpersonal awareness and she feels a need to prove how sensitive she is. Men don't. Other researchers argue that biology plays an important role and say that women are better at reading social cues because of higher oxytocin and lower testosterone levels.[27] It's probably not a single factor that explains this difference between men and women but an interplay of expectations, socialization, and biology.

Does power also play a role? If you have little power, reading other people's emotions quickly and deftly can help you keep your job. As social psychologist Carol Tavris writes, "This is not a *female* skill; it is a *self-protective* skill."[28] She points to a brilliant study done by Sara Snodgrass. Snodgrass put Harvard students in pairs to work together for an hour.[29] Sometimes two women worked together, sometimes two men, and sometimes men and women were partnered. Snodgrass varied who was in charge: half the time, a man was assigned to be the leader of the team, and half the time, a woman was assigned to lead. If women naturally have the upper hand in sensitivity, then we'd expect that women would always be stronger at reading nonverbal cues regardless of who they were partnered with and what role they played.

But that's not what happened. Instead, the subordinates were better at reading the cues of the leaders. Snodgrass found that when a man was subordinate to a woman, he was much better at reading the feelings and nonverbal cues of his female leader than she was at reading her male follower. This study suggests that when women are in charge, men can learn pretty quickly how to read the boss's impatience or interest. Snodgrass suggests that instead of calling it women's intuition, we should call it subordinates' intuition.[30]

Whatever the cause might be, the fact that women tend toward greater interpersonal sensitivity means they bring decision-making strengths to groups and teams. Woolley, the researcher who found that women are better than men at deciphering emotions from minimal facial expressions, and colleagues at MIT and Union College are trying to understand what makes for an effective team. Most of us assume that if we ask high-performing individuals to work together, we've just assembled a high-performing team. But we've all seen how a bunch of smart people in a room can still make unwise choices. To many of us, teamwork can feel like a lost cause, and we'd rather just be left alone to do a task. Woolley and her team wondered what predicted the collective intelligence, what they termed the *c factor*, of a group. The researchers gathered together people who didn't know one another, randomly assigned them to teams of two to five people, and gave them challenging problems to solve. On some of the tasks, the teams had to brainstorm new ideas and show their creative strengths. But teams also faced complex decisions: they had to make estimates based on information, decide how to allocate resources, and solve fuzzy moral quandaries. In other words, teams had to wrestle with the same kinds of decisions most of us face at work.

Woolley's findings made headlines around the world, and for good reason: a group's collective intelligence was predicted not by the average IQ score of its members or by the intelligence of the group's smartest member—the single most important factor predicting a group's intelligence was its social sensitivity. Groups that came up with the best solutions had team members who successfully read the nonverbal cues of their teammates. And that means that teams with more women reached better decisions. The collective intelligence of a group was positively correlated with the proportion of women in it. More women, better choices. Even when groups worked online, communicating through chatrooms, teams with more women showed greater collective intelligence. How could that be? Everyone knows how hard it is to read another person's tone through an e-mail message, and many of us have had our texts or e-mails misinterpreted, a light comment taken the wrong way. But women appear to have an edge in picking up on tone. Even when there's no body language or facial expressions, there's still reading between the lines.[31]

For women to improve a group's performance, of course, they need to be heard. Empathizing alone isn't enough. The problem with many real-world groups is that women are often marginalized, their contributions ignored or flattened with skepticism. In Woolley's research studies, members were mindful of how much they participated—in one experiment, each person wore a small box around his or her neck to record who spoke and who let others speak.[32] I'm not suggesting that we pass out recorders at the start of every meeting, but if we want to reap the benefits of one person's empathy, we need to listen.

If you were looking for a reason to hire more women, you just found one. We need more women on teams making crucial decisions, and not just because it's the fair thing to do. We should promote women to our top teams not only because we "value diversity," because there's a quota, or because it looks bad for the organization when there's just a token minority. We need more women contributing meaningfully to teams because the teams will arrive at more desirable destinations. This doesn't mean that solo women have an intellectual advantage over solo men. Not at all. The takeaway is that when a team has to make a choice, it needs people who are tuned in to the group dynamic as well as the factors in the decision.

We now have two things that people lump under the term *women's intuition*. There's social sensitivity—which is tuning in to others, an area where women typically excel. Amanda, the trauma surgeon, talked about developing this kind of intuitive judgment when she pushed her residents to guess, each time they stepped into a hospital room, whether a patient was doing better or worse than the day before. And then there's sensitivity to one's own deep, automatic, and inexplicable preferences, which is about tuning in to the self. Isabelle, the head chef, referred to this kind of instinct or intuition when she had to decide whether to change jobs. So we've explored two different kinds of intuition, one where women shine (reading how another person is doing) and one where women don't (reading what's best for oneself).

Is Thirty-Seven Dollars Too Much for a Bottle of Wine?

The rest of this chapter is going to focus on the more popular meaning of intuition, a concept that everyone from television personalities to inventors have embraced: the notion of going with one's inner guide. We've seen that women tend to be just as analytical and focused on data as men are, sometimes more. But can any of us, women or men, just skip the analysis? Let's say that you're not worried about what your friends or colleagues think of your choice and you won't need to justify your decision-making process with painstaking analysis. Could you reach a sound judgment if you went on gut feelings alone?

When people go with their immediate gut reactions, they're often misled by biases. Remember that intuitions aren't open to introspection, so we don't know how we arrived at a thought. We just notice that we have a sudden preference. We feel drawn in or driven away, and we have no idea if that flash of preference is rooted in something that would make us proud or embarrassed. Here's the biggest problem: Inaccurate, warped, and biased intuitions don't feel any different from accurate, informed ones. Confidence isn't even a good gauge. You can feel just as confident about a biased, incorrect judgment as you do about an unbiased one, perhaps even more so.

How much can our gut reactions mislead us? Let's explore something known as the anchoring effect. There are lots of good examples of the anchoring effect in the research, but my all-time favorite is Dan Ariely's.

Dan Ariely is a psychology professor at Duke University. In his playful and insightful bestseller *Predictably Irrational,* Ariely describes a study in which he and his colleagues at MIT and Carnegie Mellon University had students write down the last two digits of their Social Security numbers. Then they asked the students to write down the highest bids they'd be willing to make on various items, such as a bottle of red wine, a cordless keyboard, and a one-pound box of Belgian chocolates. When the bidding was over, Ariely asked his students if writing down their Social Security numbers might have influenced their bids. Of course not, they responded. That's ridiculous. But when Ariely returned to his office and analyzed the data, the pattern was clear. Students with the highest-ending Social Security numbers (80 to 99) bid much higher than students with the lowest-ending Social Security numbers (00 to 19). For instance, the high Social Security numbers were willing to pay $37.55 for that fancy bottle of red, while the low numbers were willing to pay only $11.73. In the end, the high-ending numbers bid anywhere from 216 to 346 percent more for each item. The Social Security numbers had an influence only because people had just seen them. It was a random number, but it served as an anchor, and the students unconsciously compared everything to that first value.[33]

You might be thinking, *That's too bad for those students, but I never think about my Social Security number before I pay for something.* Okay. But let's say that you're at the mall, walking through a department store. You're drawn to a display of designer jeans. You weren't planning to buy jeans today, but what the heck; you take a minute to find a pair in your size, then you hunt around for the price tag. When you see that they're $190, you think, *That much? They're just jeans,* and you quickly put them back. You decide not to look at any more clothes in this over-priced store, and you make a beeline for the main part of the mall. A smiling woman standing in front of a tea store offers you a sample cup. You try it, love it, and ask how much it costs. Only $16 for two ounces, she says. You have a surge of delight. You find yourself thinking, *That's a good deal.*

Is it? Maybe not. But with that first number, 190, as your anchor, your initial reference point, you are biased to see the second number as a steal in comparison. The conscious version of you thinks, *I'd never compare the price tag on a pair of jeans to the price tag on a bag of tea,* but the unconscious version of you sure would.

Is there a way to counteract this automatic anchoring effect? There is. Try what women do well—analyze more. Specifically, analyze the anchor. According to researchers, one strategy is to think back to the first number you saw and ask yourself, *Is there a reason that first number is inappropriate in this setting? Should I be thinking about that first number when I make this decision?* That might sound a bit artificial, so make it specific to the decision at hand; for example, in the scenario above, you'd ask yourself, *Is $190 too much to pay for tea?* The anchoring effect can be hard to avoid—that initial anchor is heavy and surprisingly hard to pull away from—but deliberately analyzing it helps you shake your strong attachment.[34]

You can't keep track of every number you see, but when you enter a new environment, especially one with a potential negotiation, pay attention to the first number you encounter, whatever it might be. If you're going to a car dealership to buy a car, let's say a Volkswagen Beetle, make a conscious mental note of the first sticker price you see. If the salesman is leading you around the lot, chances are he'll show you the most expensive Beetle first, perhaps a $35,000 loaded convertible with a built-in navigation system. You're being polite so you slide behind the driver's seat for a minute, but you say it's too much. Then he'll steer you to increasingly cheaper cars, so that the third or fourth car you see, the $27,000 hatchback with a sunroof, suddenly seems reasonable. It's a lot less than that $35,000 anchor. Your maximum budget was $24,000, but now you've adjusted your expectations. If you're test-driving that car and thinking, *You know, $27,000 sounds like a steal,* you've let yourself be influenced by a number that's no longer relevant.

To analyze your anchor, think back to the first number you saw or heard when you walked into the dealership and then ask yourself, *Would I pay $35,000 for this car? Is that number relevant?* That's the real number driving your gut feeling, the figure that's making $27,000 so attractive, so question it. You'll probably see lots of reasons why $35,000 is way too much to pay for this cheaper car — no leather seats, no rearview camera, and no touchscreen navigation system. By taking the time to think through how this car falls incredibly short of a $35,000 vehicle, you'll see that you shouldn't be swayed by that number. It can help you reclaim your original budget of $24,000, so you'll realize that even $27,000 is too much. It seems counterintuitive that paying attention to the first number would help you ignore it, but you're ensuring that you consciously analyze that first price. If you don't analyze your anchor, it will be an invisible driver of your thinking.

You might be wondering if women are uniquely seduced by the anchoring effect. No, everyone struggles with it, men and women alike. But if you take the conscious, deliberate route and question your intuitive estimate, you're less likely to make these mistakes.

When Can I Trust My Expert Intuitions?

You may be misled when you buy a car or go window-shopping at the mall, but, you insist, work is a different story. In your professional life, you face certain decisions routinely. When is it sensible to go with the first idea that pops into your head, and when should you keep analyzing your options? There is one situation when quick intuitions are something you can trust: when you're an expert who has learned pertinent skills in what's known as a "kind" environment.

Now, *kind* doesn't mean warm and supportive, like your grandmother's kitchen. It's a technical term. Robin Hogarth, an economist who used to be at the University of Chicago and is now at the Universitat Pompeu Fabra in Barcelona, Spain, says that most learning environments fall along a continuum.[35] At one end, we have kind learning environments, where feedback about one's predictions and decisions is "clear, immediate, and unbiased by the act of prediction."[36] Dan and Chip Heath give a fantastic example of a profession with a kind learning environment: weather forecasting. You predict this afternoon's weather, and feedback is quick (that day) and clear (it rains or it doesn't), which means that you can learn to recognize patterns. You can learn which cues are accurate and which ones aren't. Finally, it's unbiased — you don't change the weather just because you predict it.

Amanda was trying to create a kind learning environment for her residents in the intensive care unit. By asking doctors to guess whether a patient was better or worse moments before they looked at the patient's chart, she was ensuring that they received immediate and clear feedback on their predictions.

At the other end of the continuum, we have wicked learning environments, where feedback on predictions and decisions is slow, unclear, or actually changes because of the predictions. Teaching is a wicked learning environment on all three counts. The feedback is slow. You teach your students and you believe they understand — five kids in class had so much to say — but you don't find out if all of them learned until you see their work.[37] The feedback is also ambiguous. Students do poorly on their papers and you wonder if it's because you weren't clear in class or if they shirked the homework. And if you predict that this year's students show more potential than last year's, you'll probably spend more time giving them feedback and answering their questions, thus biasing the outcome.

James Shanteau, a psychologist at Kansas State University, carefully reviewed the evidence for different professions and found that certain careers tend toward kind environments: accountants, astronomers, mathematicians, insurance analysts, test pilots, photo interpreters, and livestock judges (yes, livestock judges). Some fields have wicked environments: college admission officers, court judges, stockbrokers, clinical psychologists, psychiatrists, and personnel selectors (hiring managers).[38] Then Shanteau identified a third group of professionals — physicians, nurses, and auditors — who perform some of their work in kind environments and some in wicked ones. Experienced nurses and physicians are likely to have robust feedback from diagnosing and treating certain illnesses, but from time to time they will encounter patterns of symptoms they've never seen. Or they'll examine a patient with symptoms that look relatively familiar but then misdiagnose and mistreat what turns out to be a different problem. In both cases, their medical intuitions are likely to lead them astray.[39]

This isn't an exhaustive list of professions, clearly, but it gives you a sense of how many careers fall into the wicked-environments category. Think of the most recent important decision you made at work. When will you get feedback on whether or not you were correct? Is that a typical turnaround time? That might help you gauge whether your environment allows you to develop reliable and valid skilled intuitions.

This is tricky. We can fool ourselves into believing that we have the expertise to trust our gut reactions. If you've been a manager for ten years and a junior colleague proposes a new floor plan for the office, you might have strong immediate doubts. Your gut reaction? *No, that won't work.* Has anyone tried it before? Actually, no. But after that first negative feeling, you're intelligent and quick enough to generate two or three plausible reasons why your junior colleague's idea is problematic. But those might not have been your real reasons for the reaction. Your strong *No* might have stemmed from an unconscious bias, maybe a general resistance to change or possibly impatience because you're too busy to consider whether or not there's a better place for the water cooler.

Perhaps the best way to determine if you can trust your skilled intuitions is the feedback test. Ask yourself honestly, *Have I regularly received immediate feedback on this kind of decision in the past?* If the answer is *I haven't made this kind of decision often enough to receive feedback,* you know you can't trust your intuitions.

You might be thinking that immediate feedback isn't important. As long as you get feedback at some point about the accuracy of your decision, you tell yourself, then you can improve your intuitions. In the meantime, you'll keep practicing the skill, and you'll get better and hone your instincts. But you're kidding yourself—immediate feedback is essential. If you don't have it, evaluating your past decisions is like trying to learn archery without being able to see the target. I know that sounds ridiculous, but imagine yourself as an archer who is happily shooting in the dark. Even if you can't see where your arrows land, you can still practice and become a more efficient archer. You'll learn to string your arrows faster. You'll build endurance so your wrist doesn't get sore, and you will undoubtedly figure out how to position your straight arm so the feathers on the arrow stop grazing your skin every time you shoot. Your body will learn many things and those will eventually come to you automatically and unconsciously. But will you become a good archer? No. If you can't see the target, if you can't immediately learn how close you were to making that bull's-eye, if you don't know whether your adjustments are actually taking you in the right direction, then what you're learning isn't going to make you more accurate. You could diligently shoot a thousand arrows, but if you don't see where they landed until a month later, you might get worse, not better.

The same is true with intuition. You might become more efficient and more comfortable making snap judgments with lots of practice, and you'll certainly grow more confident in those judgments because nothing feels new or strange anymore, but that's not the same as making better judgments. Intuitive judgments don't improve unless you see, as close to the decision as possible, whether the cues you responded to were on target.

Find Me Five Pieces of Data

Gary Klein is one of the foremost champions of intuition. He's written more than half a dozen books celebrating intuition and showing how it can be used to make better decisions. Astonishingly, though, when the business publication *McKinsey Quarterly* asked, "When can executives trust their guts?" Klein replied: "You should never trust your gut." Does that mean we should ignore our feelings and our first impressions completely? No. Gary Klein went on to explain: "You need to take your gut feeling as an important data point, but then you have to consciously and deliberately evaluate it, to see if it makes sense in this context."[40] In other words, you need to follow up on your initial intuitions with careful analysis and deliberation, just as Isabelle did when she was deciding whether or not to work at the Cordon Bleu.

So if you have an intuition, start your decision-making process with it, but be sure to go one step further: Deliberate. Deliberation involves the conscious steps you take to analyze your options, steps you could explain to someone. If you feel as though you have only one option, consider the landscape until you can generate at least one more. Do an Internet search to gather more data; look at the decision in multiple ways.

Emily, the executive vice president of sales for a rapidly growing technology company, has an approach for testing intuitions. "I have a hunch right now that there's this one area of the business that needs more attention, where our sales team could be getting better results," she explained. "So when I met with my leadership team, I told them, 'Here's my hunch. Now let's go validate whether this hunch is correct.'" Emily doesn't want her leadership team to debate the hunch or sketch out alternatives on the whiteboard or even brainstorm pitfalls in an all-hands meeting. She wants data. She told them, "Over the next week, here are the five things I want you to go look up and analyze. I'm pretty certain my hypothesis is going to be right, but we need to gather the facts." Several people in the room thought she was onto something, and several others thought her idea was overstated, but they all agreed they needed data.

Might the team gather only the information that confirms what the most powerful person in the room wants to hear? It's possible, and Emily recognizes that people often use data to confirm and spin the tidy story they want. But she uses two strategies to guard against that happening. First, the group, not Emily, generates the questions. "We all agree on which five questions need to be asked and answered to con-firm or dispel the hunch." Notice that she's not challenging the team to predict the future; she's asking, "What evidence would convince us either way?" Second, the group starts with questions, not data, because if you begin with a single data point, you can justify almost anything. To see how easy it is to milk a single data point, let's take an example from a different industry. Say you read that Americans spend over $5.4 billion a year on ice cream.[41] You've always wanted to open your own business, ice cream is your favorite indulgence, and you make a pretty mean salted-caramel swirl. Suddenly everything clicks. That one piece of information you've picked up fits your personal passion, and now you're thinking, *If I bought an ice cream truck, I could pay the bills while doing a job I love.* But just about anything can click into place around one piece of data. You can make a much more informed decision if you start with questions that generate at least two pieces of information that you can put side by side and compare, questions such as "What do the cities with the top ice cream sales have in common?" and "What percentage of that $5.4 billion is ice cream truck sales and what percentage is grocery-store sales?" and "How many hours a week do most food-truck owners work their first and second years?" Now you're not justifying, you're analyzing.

Does Emily's approach work? Her tech company's sales have grown at least 70 percent a year for three years running. In sales, anything over 30 percent is called hypergrowth. Emily is quick to clarify that her "find me five pieces of data" approach isn't the sole reason for the business's enviable success — brilliant scientists, an enthusiastic sales force, an innovative marketing team, not to mention a timely product, are the backbone — but her mindset ensures that her colleagues don't impulsively pursue an idea based only on her Spidey-sense. Intuition is a piece of how they make decisions; it's not the beginning and end of the process.

Knowing it's a good time to put on your analytical thinking cap doesn't mean it's an easy thing to do. Sometimes intuitions can be strong and hard to ignore. Thankfully, researchers have discovered some ingenious ways to help us switch gears from an intuitive mode to a more analytical mode. If you're feeling impulsive and need to slow down and analyze your options more carefully, furrow your brow for thirty seconds. It's simple, it's odd, but it works. Furrowing your brow puts you in a critical mood, a mood where you're less spontaneous and more likely to pick apart the details of any decision, including your own.[42] Sad moods also put people in a more analytical mode, and recalling a sad memory can give you some distance from a strong initial preference.[43]

You don't have to sabotage your mood, however. Some people find they just need to say to themselves, *Think carefully,* or *Evaluate each reason,* and they become more analytical. If you're trying to get the members of a group to move away from their gut reactions to a proposal, try telling them: "I need you to be highly confident in whatever choice you make." The pressure to make the best choice possible helps people take a more thoughtful and studied approach. But some kinds of pressure backfire. Giving a group a tight deadline or announcing "I need a decision as soon as possible" makes people more likely to fall back on their intuitions. It's better to say, "Take the time you need. If I can get a decision today, great, but I'd rather have a good decision than a fast one." Or "I don't want to rush this. I want us to spend a full forty minutes discussing our options before we try to decide."

For Anyone Who Believes in Deciding with Your Heart

If you're someone who likes the idea that women have more powerful internal compasses than men, that women can tap into their hearts or their guts to make enlightened choices, you may resist the notion that women aren't more intuitive. But recent research shows that how you think you make your decisions shapes the quality of your judgment.[44]

Adam Fetterman and Michael Robinson, two psychologists at North Dakota State University, conducted a series of studies to see how the choices of people who said they decided with their heads compared with the choices of those who said they decided with their hearts. The findings were clear — thinkers had the upper hand. University students who saw themselves as deciding with their heads (not their hearts) thought more deeply about questions in the moment, made choices that benefited a greater number of people when faced with a moral dilemma, and earned better grades overall in their courses.

The issue of grades might raise the cause-and-effect alarm. Perhaps the grades came before the head-or-heart mindset. Perhaps students who did well in their courses began to see themselves as studious types, as people who really knew how to use their heads, while students who weren't at the top of the class reassured themselves, *Well, my talents lie elsewhere.*

It could very well be that grades shaped how people thought of their strengths. But the researchers showed that people could easily switch their metaphors for decision-making, and when they did, they solved problems differently. Fetterman and Robinson asked half the people in the study to rest their index fingers on their temples, literally pointing at their heads, while answering the test questions. Both men and women answered more analytically, less emotionally, when they pointed at their heads. Tapping one's temple may seem rather contrived, but it's not as strange as it sounds; tell someone sitting at a desk to think hard about a problem and there's a decent chance he'll rub his forehead or prop his chin in his palm. But when people were told to rest their index fingers on their chests, men and women alike took a more emotional, less rational approach to deciding how to resolve a moral dilemma. What's fascinating is how such a small change in position can elicit such a big change in thinking strategies.

This is only one set of research studies, and some researchers disagree with the larger premise behind this work—namely, that one's thoughts can be influenced by one's body.[45] But perhaps the most important lesson from this line of research isn't that pointing is important; it's that you can move fluidly between decision-making styles. Even if you've always sworn by the "I follow my heart" motto, you can try saying, "I also know how to use my head," the next time you face a hard decision. See what new solutions spring to mind.

But My Gut, That's Me

Perhaps you're still not convinced. You might be running into the issue of authenticity. Some women and men feel their decisions are more authentic, more truly reflective of who they are, if they follow their intuitions. As one of my friends puts it, "My head's full of what other people tell me I should do. But my gut, that's me."

I can certainly understand what she's saying. When you sit down to analyze your options, other people's preferences can blur your own. If you're deciding whether to go back to school to pursue a new career, you might think about how your current boss will react, what your partner will think of your drop in income, or whether your mother, who was so proud when you chose your current career, will be disappointed. You consider the impact of the change on people who count on you, be it your kids at home or your team at work or both. With all of those competing interests, it can be hard to pluck out what's truly important to you. You can fill pages and pages with notes, but nothing stands out. You find yourself hoping for a strong intuition that you can't explain, trusting that any gut feeling will be "you" shining through this tangled mess.

If you're concerned that your head will be too filled with what other people want, if the reason you're drawn to go with your gut is that it helps you untangle your bedrock needs from everyone else's, there is something else you can try, something we'll call a look-back. It's based on a strategy we'll learn more about in the chapter on risk-taking.[46] First, imagine that it's one year after you've made your decision, then complete this sentence: "Looking back, I'm so glad that I . . ." Or this one: "Looking back, one of the most important choices I made this year was to . . ." These look-backs help you identify what's most important, what your biggest priority is. Another favorite look-back that my husband and I both use is "If I hadn't _____ this year, I would really regret it." One evening in 2013, we were playing with that sentence and I said, "If I hadn't gone to Paris this year, I would really regret it." He put his arm around me and said, "What's in Paris, besides you and me?" And when we did go, I met Isabelle and this whole story began.

Why is it helpful to imagine that it's a year from now and you're looking back? We'll unpack that question when we talk about risk-taking, but in the meantime, consider this: Is it easier to reflect on the past or predict the future?

Note how different a look-back is from making a list of pros and cons. A look-back is retrospective, but there's another insight that's easy to miss. All too often, decision-makers generate lists of pros and cons to help them decide an issue, and they fill entire pages hoping that with each new item, they're getting closer to clarity. But longer isn't better. As Chip and Dan Heath observe in their delightful book *Decisive,* lists of pros and cons often "kick up so much dust that we can't see the way forward."[47]

The Problem with Typical Job Interviews

Let's take a look at one place where intuitions have too much sway in professional lives: hiring decisions. Chances are that you'll be in a position to hire someone at some point, whether it's a contractor for your home or an employee for your office. So imagine you need to hire a new administrative assistant. You have the money, you have cleared time on your schedule to interview applicants, and you have meetings scheduled with four outstanding candidates. What's the most common error people make at this point?

The mistake is to sit down with each of them and just chat. Prospective employers do this frequently, assuming they'll get a sense of the candidates by letting them talk about themselves and their relevant experiences. This kind of unstructured interview is one of the most common steps in the hiring process, but it's also one of the worst.[48] Even a simple, web-based intelligence test is a better predictor of on-the-job performance than the typical interview.[49] The problem isn't that potential employers are bad at reading people; the problem is that what they're reading is probably a poor predictor of job performance. A half hour is all the time most of us need to develop strong feelings about a job candidate and start thinking, *She reminds me of myself twenty years ago!* or *She's okay, but I just don't see anything special in her.* It's one of the reasons that hiring managers love unstructured interviews. People believe, or perhaps they just hope, that if they talk with the candidates awhile, asking them to describe their previous jobs and where they see themselves in five years, they'll pick up on something important, some essence or quality that makes one person stand out as someone with real potential or another person reveal himself as a real problem.[50] These first impressions, those sudden bubbles of like and dislike, might help you decide whether you'd like to sit next to that person for two hours at a dinner party, but they're a poor gauge of whether you'll be impressed with his ability to manage a team or get a project in under budget.

Some people might agree that a casual interview over coffee is not the best way to evaluate a candidate, but it can't hurt, can it? Surprisingly, it can. Researchers have found that once an interviewer forms an impression of a job candidate, even if it's quite arbitrary, he often can't let go of it. If the candidate asks if the cappuccino can be made with hemp milk or if she excuses herself to take a phone call, the interviewer might feel something is slightly off. He can't put a finger on it, but he gives too much weight to that irrelevant cue and ignores the real diagnostic information, such as the candidate's stellar past job performance. Or take a smooth candidate who is charming and witty but not very qualified. When the hour interview flies by, the manager rationalizes that the key qualifications listed for the position aren't really all that key. Psychologists call it the dilution effect, and they see it again and again in job interviews. An irrelevant observation dilutes what's really important. You hired one person because she seemed so confident and worldly in the interview, and six months later, you find yourself frustrated because she's regularly asking for time off to travel.

You might be thinking, *Well, I just need more practice and then I'll know what to look for in these conversations and I'll be able to trust my gut.* But research shows that even people who conduct interviews professionally, who spend all day screening candidates for positions, are terrible at predicting a person's future performance based on an interview.[51] It's easy to get wrapped up in liking or disliking people, and try as you might, those feelings can be hard to ignore. As we learned earlier, intuitions can be trained only when there's immediate and clear feedback. Immediate, as in a few minutes later, or at least the same day. Typically, weeks go by between the interview and the employee's first day on the job, and no one accomplishes much on that first day anyhow. For complex professions, it's another month or more before you can assess a new person's performance, and at that point, your gut feelings are long gone. You can't retrain how you developed your first impressions that far out. Fortunately, you can still train your conscious reactions. You can review the questions you asked in the interview and pinpoint which ones were too vague, and you can decide you need to call at least three references. You can improve how you deliberate. But with that much lag time, you can't improve your unconscious, intuitive reactions.

So if unstructured job interviews are so problematic, what's the alternative? One of the best strategies is to ask people to complete a work sample, which gives the candidates a chance to simulate the kind of work they'll be doing. If you're hiring someone as a consultant, bring in a sample client with a problem and observe quietly on the side while the two of them work together. If you're hiring someone to design a website, give each applicant a computer with the right software and an hour alone in a room. Obviously, candidates aren't going to be able to create a full working website, but it will give you a chance to compare how they approach the task.

I've used this technique several times, most recently to hire a research assistant to help me with this book. I had three promising candidates, all of them still in college, all of them with strong résumés. I met with each of them individually for an hour and explained that rather than conducting a typical interview, I was going to have them do a sample research assignment. I gave them the type of project they would be doing on a regular basis if they were chosen for the position: I asked them to find an example of a woman making a particular kind of decision. It could be an American executive, a German politician, a Brazilian athlete—anyone they wanted.

I asked that they spend a maximum of seven hours, including the write-up, on the project and explained that I had chosen a very difficult topic, so it was possible that they wouldn't find anything in the allotted time. That was okay. I told them to stop searching at five hours and begin documenting their thinking and the search process, including any sources used. Then we spent most of the rest of our hour together discussing the topic they had to research. I said that I would pay them a set amount for their time, asked them to e-mail me their reports in ten days, and told them to contact me if they had any questions.

No questions came in. Ten days later, I received three e-mails. The first person I interviewed apologized and said she was withdrawing her application. She'd spent four hours looking and couldn't find anything and realized this wouldn't be a good fit for her. The second applicant had found an example, and although it was interesting, it didn't illustrate the concept we'd discussed. She didn't acknowledge that in her e-mail. The third person I interviewed also couldn't find anything, but she gave a detailed, three-page analysis of the different sources she'd considered. She explained where she'd looked and why none of the examples she'd found quite fit.

So based on those results, who would you hire? I chose the third applicant. She wasn't able to find an example but her thinking was clever, she'd documented everything beautifully, and she'd considered some angles I hadn't. And she'd clearly followed my directions. But here's what is most revealing: if I had made my decision right after the hourlong conversations, I would have offered the job to the first candidate, the woman who ultimately withdrew. Based on our time together, I felt the best about her. Of course, I could have consciously justified that decision — I would have pointed to the fact that she asked the most insightful questions. But is that the real reason I was drawn to her? I can imagine a dozen unconscious things that might have triggered my reaction. Perhaps she reminded me of a past star student or some long-forgotten babysitter. She might have pronounced my name correctly on the first try, or maybe she was wearing an outfit that struck a chord with me. Who knows? The third applicant, the woman I did hire based on the quality of her research, was very pleasant in the interview, smiling and nodding, but she had no questions for me. She was quiet. After she left that initial meeting, my gut reaction was *Nice, but maybe too shy.* I'm not sure what I was reacting to, but I wouldn't have hired her based on our conversation alone. But in the simulated research process, her work sample was the most impressive. And research was, after all, why I was hiring her.

Did I hire the right candidate for the job? I am so impressed with Nikki. She's fast, reliable, and extremely clever in helping me find sources. She remained quiet in our first meeting or two, but gradually she came out of her shell, and by the third month, she practically led our meetings. And she regularly stumped me with her quick observations and insightful questions. The work sample revealed a strong candidate I would otherwise have passed over.

Would the first person have panned out? Maybe, but she found the research, the very task I needed her to do, a poor fit. Maybe it was discouraging for her when she discovered how challenging the task would be and ran into all those dead ends. She was a straight-A student, and many top students aren't accustomed to failure. It was good for both of us to find out ahead of time that this wasn't the job for her.

Maybe a simulation isn't an option for you. If you're limited to a traditional interview, and sometimes you will be, ask questions that elicit factual information, such as "How many administrative assistants [or bosses] have you had in the past five years?" and "How many grants did you work on last year?," rather than subjective, skewed queries such as "What is your greatest strength and your greatest weakness?"[52] If you learn that the prospective employee in front of you has had five bosses in as many years, you need to ask more questions to figure out if this person was simply caught in an office with high turnover or if she has a hard time with authority figures or some other broader issue. If you're hiring someone who will be submitting grant proposals, you need to know if his expertise dates back to the fax-machine era or if he's going to be able to teach you a few things.

. . .

To make great decisions, we need to abandon one notion of women's intuition and embrace the other. The idea that women know the right course of action simply by looking inward, that the right decision will be there, waiting, fully formed, isn't helpful. Women might think they're tapping into their priorities, but they might just be tapping into their biases. If they relinquish that notion, they'll ultimately find themselves making decisions that align with their priorities. Women are often more analytical than they want to admit, but once they do admit it, that acceptance frees them to strengthen the decision-making process even more.

But what about the idea that women are often more attuned to the people around them? This is the kind of women's intuition that we can and should embrace. Contrary to what sitcoms suggest, women aren't mind readers. But women are more adept at picking up on the emotional cues of the people around them, and when they do, their social acumen can raise the intelligence and good judgment of the entire group.

Chapter 1 at a Glance: Making Sense of Women's Intuition

THINGS TO REMEMBER

1. Women's intuition is seen as a powerful and unique way that women make choices.
2. Scientists find that expert intuitions, whether made by men or women, are fast, holistic, and unconscious, and, to be useful, they require a lot of practice with clear feedback.
   - Example: Doctors quickly guessing whether a patient is better or worse before they review the patient's chart
3. Many people think they have good instincts, but the reality is they often haven't received immediate or clear feedback on

their guesses, so their gut feelings may not be as accurate as they'd like to believe.

4. Women tend to rely on thoughtful, deliberate, and conscious analysis just as often, if not somewhat more often, as men.

5. Compared to men, women have learned to be more accurate at deciphering emotions from others' facial expressions and body language.

   - Because women pay more attention to how different group members are reacting, increasing the number of women on a team and seeking their input can raise the team's collective intelligence.

   - It's not really women's intuition; it should be called subordinates' intuition.

THINGS TO DO

1. When you're deciding how much to pay for something, recognize the anchor effect.
   - Example: Ariely's students bidding on bottles of wine

2. Use your intuitions as a starting point, then hunt down data.

3. When you face a hard decision, remind yourself that you're using your head, because that mindset leads to clearer thinking than telling yourself you're following your heart or going with your gut.

4. Try a look-back when you need to untangle your emotions from everyone else's.

5. When you're hiring, don't use a typical unstructured job interview where you get to know the candidate.
   - It can lead you to follow unreliable and unconscious likes and dislikes.
   - Ask candidates to do the kind of work they'd actually do on the job and hire based on performance.

2

The Decisiveness Dilemma

WHEN I SAT DOWN in Diana's office, she picked up a remote and muted the already quiet television. "I always watch the news all day," she explained. "In case there's a breaking story." Diana wasn't a newspaper reporter or a CNN anchor. Diana was a police chief.

Most of us couldn't say "I watch the news all day" and maintain any authority in the office, but no one would suggest that Diana was slacking. As a commander of a city police force that has over five hundred full-time officers, one of only eighty-nine police departments in the United States that large, she needed to know what was happening locally and nationally so she could make decisions that were incisive, contextualized, and informed.[1]

Television shows like *Law and Order* and *CSI* would have us believe that women are commonplace in law enforcement. Sure, men outnumber women there, but for every two or three men on these programs, we see at least one woman detective. Even comedy mockumentaries like *Reno 911!* show a nearly even split of bumbling male and female cops. But in the real world, women are a rarity in this field. Females make up 15 percent of all police officers in the United States, which means that for every one woman there are almost eight men.[2]

That means that if you're a woman officer, you'll probably have a male partner for most of your career. Forget *Cagney and Lacey*. And as of 2013, one study of 1,550 police departments in the United States showed that over 150 agencies didn't have a single female officer.[3]

But I wasn't there to ask Diana what she thought of TV cop shows or the uneven gender representation in law enforcement. I'd come to discuss how male and female officers make decisions in the line of duty. At first, Diana insisted there was no difference. "All of our officers receive identical training," she explained. "It doesn't matter whether you're a man or a woman; you're taught to react the same way."

There was no doubt in my mind that all officers were taught the same carefully planned procedures, yet I still wondered how these were used in the field. In some circumstances, there had to be room for an officer's own judgment, some degree of choice. When there was more than one potentially effective way to handle a situation, did men lean one way and women another? Diana gave this question some careful thought. "Every officer is different, just as every human being is different," she finally said. "But there are patterns you notice if you do this long enough. Let's say an officer arrives on the scene and two people are going at it, yelling at each other and threatening to get violent. Many guys are going to run in head-first and try to get the situation under control. Their decisions are action-oriented. They might say, 'What's going on here?'" She barked the question in an authoritative tone. I pulled back in surprise.

Women officers, Diana said, asked the same question, but they often asked in a more compassionate tone, a manner that encouraged a conversation. "A woman might walk in and say, 'Hi. You know what happened? What's going on?'" Now Diana sounded like she was talking to a bystander or someone who'd been wronged. All the while, Diana said, women would be analyzing the situation. "A woman is going to be thinking, *Who's the biggest person in the room? Can I take him down if I need to? And most of all, what can I do to de-escalate this?*

"But depending on the male officer she's partnered with," Diana said, "that approach doesn't always sit so well. If he doesn't know her, he can think she's stalling. It might seem like she's not moving toward a solution, like she's being too hesitant." Like she's taking too long to decide. A person might misinterpret this measured approach because an intense situation seems to warrant an equally intense response. Her partner might be thinking, *Now is not the time for nice.*

I started looking into decisiveness and whether people see one sex as procrastinating a bit more on important decisions, but in almost every conversation I had with women about making firm, swift decisions, a second issue came up — taking care of other people. I thought I was asking a simple question — "In your experience, is one sex more decisive than the other?" — but no one, including Diana, let me off that easy. When I tugged on decisiveness, stories about inclusion and compassion came with it.

This chapter will still ask whether one sex is more decisive. Do men decide more quickly and with less internal struggle, as many people seem to believe? But we'll also look at the complex tug of war that women face between being decisive and being responsive to others. It's hard to be seen as both. What price do women pay when they include rather than insist?

We'll also address a question that's going to be important throughout this book: What happens to women when their colleagues suggest, perhaps very subtly, that women aren't as good as men at making decisions? What happens in an environment where the cues strongly encourage women to use one process for reaching a decision while great leaders use another? Do these environments change how women make decisions? And what can women do to stay savvy and beat back the stereotypes?

The Only Time You Hear the Word *Prerogative*

You've probably heard the phrase *It's a woman's prerogative to change her mind*. It seems harmless and old-fashioned, like something my grandfather might have said, and it's often used in jest. I heard a grocery-store clerk say it at the checkout stand when the person in front of me decided against a third jar of spaghetti sauce. I've heard women jokingly say it in meetings when they want to retract a bad suggestion they've made, and believe it or not, I once heard a father shake his head and say it when his daughter, at most three years old, dropped one doll to pick up another.

The phrase *woman's prerogative* is indeed an old one, and if women knew its origin, most would hesitate to use it. It's not a testament to women's power or their special privilege; it's a testament to how little power women once had in a very specific situation. In England in the 1800s, if a man proposed marriage to a woman and then changed his mind, she could sue him for breach of promise, because a marriage proposal was taken as a legally binding contract.[4] He proposes and she's taken. Nineteenth-century society treated a woman left at the altar as damaged goods (according to historians, some men at the time would have assumed that she'd lost her virginity to her betrothed if it was close enough to the wedding date, and besides, what was wrong with her that would make him leave her like that?). These rejected women often found it hard to find another match. In recognition of this problem, the law allowed women to sue men for broken engagements as a way to compensate them for ruined reputations and lack of future prospects; in the 1850s, the woman was awarded around £390 (which in 2014 would have been comparable to approximately £530,000, or $839,000).[5] In other words, if a man changed his mind, he might be forced to pay. A woman, by contrast, could change her mind and break off an engagement without being dragged to court. This is where the phrase originates; she could call off the engagement without legal or financial repercussions, but he couldn't. The law typically ruled in a woman's favor in such suits too. Why? Men held all the power and suffered little from a broken engagement. If a woman called off the betrothal, the man would have no problem finding a wife, while the woman, even though she was the one who had changed her mind, would be shunned by society (remember, her virginity was in question). This was a woman's privilege or prerogative: she could get out of a terrible engagement, but if she did, her reputation and her marriage prospects would be ruined. If the man called things off, he could pay a hefty fee and get on with his successful life.

Journalists and pundits continue to dredge up the *woman's prerogative* phrase when powerful women update their views based on new developments. When Barbara Bush said that she would support her son Jeb Bush in his bid for the U.S. presidency, she retracted a statement she'd made two years earlier about not wanting another Bush in the White House.[6] Chris Matthews, on MSNBC, criticized Hillary Clinton because she wouldn't admit that she regretted voting for the Iraq war when she was a senator.[7] In both cases, journalists wagged their fingers at these individuals but then made the "woman's prerogative" remark as though these women reflected the flaky nature of women everywhere. When powerful men such as Barack Obama, David Cameron, and Pope Francis change their minds, they aren't a source of shame to men in general. Journalists might accuse these individuals of being hypocrites but they rarely extend the criticism to everyone with a Y chromosome. The notion that men are more decisive than women might be old-fashioned, but it's not passé and it maintains a strong hold on society. A 2015 survey had Americans go down a list of positive personal qualities and state for each item whether they thought it was more true of men or of women.[8] Respondents thought that women were more likely to be compassionate and organized, but men were believed to be decisive. Another survey found that being decisive was one of the ten traits that people believed "typical American men" were likely to possess, but the same group didn't think *decisive* described typical American women at all. *Warm, kind, friendly,* and *patient* were all highly fitting adjectives for women, the respondents felt, but *decisive* fell near the bottom of the list of forty-three terms.[9] Does that mean everyone thinks men are more decisive than women? No, but it does mean that when most people think about the favorable qualities that men possess, one of the first words that comes to mind is *decisive*, but they have to sit with pen and paper for a long time before the word *decisive* comes to mind for women.

Let's clarify what most people mean when they say that someone is decisive. Nearly every scientist who studies this quality has a different view of it, but one simple definition that fits how most people commonly use the word is that a decisive person has a "bias for action." That's the phrase that Tom Peters and Robert Waterman introduced in their business bestseller *In Search of Excellence.*[10] Peters and Waterman studied forty-three successful companies in the United States and found that decisiveness, or a bias for action, was one of the principles that highly profitable, well-managed companies had in common. Take a task force. How long should a task force study a problem before it suggests an action? Peters and Waterman found that the best-run companies were more likely to have a "swarm of task forces that lasted five days," whereas the norm in an industry might be to have a single committee study a problem for eighteen months.[11]

If decisiveness is a bias for action, what's the opposite of decisiveness? A bias for thought? Yes and no. Researchers make a distinction between indecision, which is seen as a normal and healthy thing, and indecisiveness, which is seen as a problem. Indecision is considered a temporary stage that adults often go through at some point when they face a significant, potentially life-changing choice.[12] If you're deciding whether to go back to school or whether you should lay off an entire division, you might go through a period of indecision commensurate to the high stakes.

But indecisiveness is a personality trait, not a stage in a problem you're sorting through. Indecisiveness is "a chronic inability to make decisions in various contexts and situations."[13] If you're indecisive, you have difficulty making even small choices. You consistently have trouble choosing what to order in a restaurant, where to sit in a meeting, and which book to take on a trip. (If you read that last sentence and thought, *Those are all hard choices,* you might very well struggle with indecisiveness.) So whereas decisiveness is a bias for action, indecisiveness is a bias away from choosing. And that can be one perception of women — that they have a chronic bias away from choosing or away from acting.

We Want a Leader Who Decides and Doesn't Look Back

Is it important to be seen as decisive? Politicians think it's crucial. For years following the September 11, 2001, terrorist attacks, Republicans represented U.S. president George W. Bush as a highly decisive individual, one who made immediate, gut-level decisions and stuck to them despite criticism and opposition. Bush's supporters cited this as one of the reasons the country was safer under his leadership, and his comment "I'm the decider" is one that many associate with his presidency.[14] Contrast that with how Americans have scorned indecisiveness in recent decades; in the 2004 presidential election, Republicans shamed the Democrat nominee, John Kerry, for being a "flip-flopper"; they were hoping to paint him as a poor leader and turn voters away from him and toward Bush. And it worked. In that election, perceptions of decisiveness were crucial.[15] One study found that voters who saw George W. Bush as the more decisive leader cast their ballots for Bush, and voters who saw John Kerry as more decisive gave Kerry their votes.

Americans still prize decisiveness today, according to a 2015 study that asked "What makes a good leader?" A team of Pew Foundation researchers asked twenty-eight hundred American adults in different roles and occupations what they considered absolutely essential qualities in a leader. Honesty ranked number one, with 84 percent of respondents saying that truthfulness was essential, but decisiveness was close on honesty's heels, with 80 percent of respondents saying that effective leaders must also be decisive.

Did people think that men and women differed on this key leadership quality? According to this study, 62 percent thought that men and women were equally decisive, but 27 percent thought that men were more decisive and that this made men better leaders. The good news in that statistic is that the majority of the people thought men and women were equally decisive. That's progress. But here's the concerning part for women who have been following the advice of many popular business books and ratcheting up their ambitions: of all the leadership qualities Pew studied, decisiveness, not ambition, was where respondents said women lagged farthest behind men. Twenty percent of the people surveyed thought that men were more ambitious than women, but 25 percent thought that men were more decisive. It's a tiny difference, but it signals that women can't be focused solely on showing that they *want* to lead; they also have to show that they're decisive enough to lead.

It's not just Americans who revere decisiveness in their leaders and abhor sudden changes in opinion. In the United Kingdom, government officials are criticized for "U-turns," and in Australia and New Zealand, people dislike politicians who "backflip." And in one broad study of twenty-two countries in Europe, all but one nation ranked decisiveness as one of the top five qualities of an outstanding leader. (France stood alone in its acceptance of leaders who took their time deciding.)[16]

Enough with the Meetings Already

Does everyone see women as the less decisive sex? Thankfully, no. When I spoke to women at the highest positions in their organizations, they said they were often more eager to decide than the men in the room. Kat is a former CEO of a tech company who said, "I usually wanted to be ninety-five percent sure of a decision before I gave it the go-ahead. But some of the men I worked with? They wanted to carry it out to four nines" (meaning, she kindly explained, they wanted to be 99.99 percent certain). Diana, the police chief, told me that when it comes to making important policy decisions that affect the entire police force, "Men plan to have a meeting to have another meeting to have another meeting. Meanwhile, I'm thinking, *We need to stop having all these damn meetings. Let's just make a decision already.* Come to think of it," Diana added, "I said something like that in a meeting yesterday."

This may speak to the kind of women who make it to executive levels in male-dominated environments. Diana is the first female police chief in her city in over a decade. Her department is far more progressive than most. As of 2013, less than 1 percent of all police chiefs and sheriffs in the United States were women, which means that if five hundred police chiefs were gathered in a room, you'd expect only four of them to be women.[17] Diana walks a path that has been traveled by very few other women. And if you're a woman in a profession normally associated with men, it may be easier to rise to a leadership role if you demonstrate, loud and clear, that you can, and will, push for a decision.

We put a premium on decisiveness in our leaders, so when people do believe that women are lagging behind in making crisp, clear decisions, that perception translates into a number of problems. Sometimes women's decisions are dismissed as unpredictable. In a recent article in the *Harvard Business Review* on the different processes men and women use to reach a decision, one unnamed male partner in a large firm admitted, "I think sometimes women are so much more difficult, even fickle, in business dealings."[18] Or women, who are seen as both relationship-focused and prone to change their minds, are heavily lobbied while men are left alone to weigh the options. Nina, a software engineer and product manager, observed that in her division, there was a belief that women decided based on the "last person who touched it" principle. I'd never heard that phrase before and asked what it meant. She explained that when a female manager in her group had an important decision to make on a project, people would keep swinging by her office on the day the decision had to be made. The assumption, Nina said, was that a woman would go with the last suggestion she heard. Male managers? "No lines outside of their doors," she told me. "People figure, 'What's the point? He's going to do what he's going to do. We'll find out once the decision is made.'"

I found Nina's perspective fascinating. I'm not saying that the "last person who touched it" principle is common in the workplace — it could be unique to this company, or even this division — but it illustrates a larger issue. It fits the message we've heard so far, which is that people see men as biased for decisive action but expectations for women are mixed, at best. Does this team think women are ridiculously capricious or simply open to new arguments? Is this a hidden compliment — that women care more about what the team thinks? Or is the implication that it's easy to redirect women, that they have no minds of their own? One of the best ways I've found to understand this story and the thorny decisiveness issue is to think about secret agents and mother ducks.

Secret Agents and Mother Ducks

In general, we expect very different things of men than we do of women. If you meet a guy at a friend's backyard barbecue, you might not be surprised to hear he has big ambitions, even if he's wearing swim trunks and has a toddler swinging from one arm. The cues in front of you tell you that he's family-oriented, easygoing, perhaps even playful, but the ingrained social expectation is that he'll still have a career and set his sights high. Researchers find that we expect men to be achievement-oriented and have an independent vision for themselves. If he works for a firm, you'll nod as he tells you that he's on the fast track to becoming a partner. If he works for himself, you might ask whether he plans to remain a one-man operation or expand someday. Even as you watch him lean over to help his daughter tie her shoes, you assume, without realizing it, he's someone who also leans ambitiously into his career. A book titled *Lean In* wouldn't have been a compelling career guide for men because that's what we expect them to do.

Scientists say that we expect men to be agentic — to drive their own agendas, to set their own goals and, ultimately, achieve them.[19] *Agentic* is a technical term, but I like to reimagine it as describing a secret agent. You've seen this kind of gender role in a dozen movies. Our hero is waiting on a rooftop for instructions, and the person at headquarters finally gives the command: "James, we need you to do X, Y, and Z, in that order. James?" The camera cuts back to the roof, and the audience sees an earpiece, presumably his earpiece, abandoned on the ground.

He's off doing whatever he's decided is best, and moviegoers are both anxious and thrilled, bothered that he's ignoring the plan but trusting that he's pegged exactly what's needed.

When it comes to making judgments and decisions, we expect men to be like secret agents: competent, independent, forceful, and, most of all, decisive. Quick to decide, quick to act. We also aren't surprised when the plan they follow is their own. This doesn't mean we like those qualities in men or that we want the male decision-makers around us to behave this way, but when we see it play out like this, we're not stunned.

When it comes to making decisions, men are expected to behave like secret agents, so how do we expect women to behave? I hate to say it, but this time, I picture mother ducks. Scientists say we expect women to be communal, to focus on their communities, on building and maintaining relationships. We expect women to be more like mother ducks when they make choices, paying attention first and foremost to the needs of those around them. If you picture a mother duck and her brood crossing the street, you know she's going to make sure everyone gets to the other side. Proceeding ahead of everyone is fine as long as she's simply clearing the way, but we assume that she has the group's interests in mind and that she's keeping an eye out for any stragglers. Do we think that her decisions center on her own independent agenda? That would be a terrible mother duck. No, we expect her decisions to reflect the helpful, concerned, and sympathetic role she plays for those who depend on her.

What does this mean for women and decision-making? It means that we expect women to focus on something larger than themselves, whether it's the company, the team, or the family. Let's imagine you meet a woman at that same barbecue, and she's wearing jeans and showing a toddler how to blow bubbles, and she says she's just been promoted to VP. You ask her (but not him) how she's juggling career

and family. You ask her (but not him) how she does it all. You can easily imagine him as the secret agent, but you can't shake your sense of
her as the mother duck.

We see the secret-agent and mother-duck mindsets all the time
among business executives. In a 2009 study of almost three hundred
senior business managers, the common perception among executives
within two reporting levels of the CEO was characterized as "Women
take care, and men take charge."[20] Most of these senior managers, including female executives, saw male employees taking decisive action,
taking control of a situation, but they saw women as nurturing and
supporting others.

These aren't perfect analogies that capture every complexity that we
expect of men and women. They especially fall short around issues of
receiving credit and recognition. Secret agents operate behind the
scenes, doing the impossible and then disappearing before anyone
knows who was responsible, but that's not what we expect of most
men. We don't, as a society, require men to be quiet about their accomplishments. And likewise, a mother duck towers above her ducklings
and stands out as the one who's in charge, whereas tales abound about
women leaders being mistaken for the person who gets the coffee. So
this analogy doesn't work if we're asking who's going to be recognized
in the room.

But it does work if we're thinking about how men and women decide, whether they do it independently or try to be in tune with others.
We expect him to move ahead if he has a better idea, but we expect her
to be collaborative, to take everyone's needs into account. We see it in
Diana's story, where male officers are inclined to take charge while female officers are inclined to take care. We see it in Nina's story, with
employees lobbying female managers on the day of a crucial decision
but leaving their male managers alone. We saw it in the introduction:
Plenty of journalists criticized Marissa Mayer for ending Yahoo's work-

from-home policy (Wasn't she thinking of all the working mothers at Yahoo? Shouldn't a female CEO be more family-friendly? And what precedent was she setting for the industry; had she thought of the bigger picture?), but few castigated Hubert Joly when he made the same decision for Best Buy a week later. He wasn't supposed to watch out for every last duckling, but she was.

Does that mean that men can't be communal and that women can't be agentic? They can be, but as we'll see throughout this book, men and women often pay a penalty when they adopt the gender role considered typical of the opposite sex. Pundits and journalists are especially quick to criticize women leaders who approach decisions like secret agents, going full steam ahead with the choices they think are best.

Who Has the Edge in Decisiveness?

Are men, by nature, truly more decisive than women? Researchers still debate the question, but the current answer seems to be "probably not."

Most scientists who study decisiveness rely on adults' ratings of their own tendencies. If you took one of these surveys, you'd answer twenty to twenty-five questions about how you approach important decisions. For instance, to what degree do you double-check your facts before you make a decision? How often do you make important decisions without consulting other people? How often do you postpone making decisions until the last possible moment?[21]

When researchers tally answers to these questions, men and women earn pretty similar scores.[22] The data suggest that everyone finds it hard to make some decisions some of the time, but most of us, men and women alike, move through life smoothly enough, readily making the decisions that need to be made. This gender balance in decisiveness isn't limited to the United States. In nations as various as Turkey,

China, Canada, Australia, Japan, and New Zealand, men from each country look just as decisive as women from that same country.[23] One culture might be more decisive than another, but within a culture, men and women seem balanced. For instance, one study found that Japanese adults were less decisive than American or Chinese adults, but Japanese men and women showed little difference in how swiftly or easily they chose.[24]

Some researchers do report that men are more decisive, but this comes with an important caveat: most of these studies were done on two special populations, specifically adolescents between the ages of twelve and eighteen and adults who exhibit signs of obsessive-compulsive behavior, such as repetitive hand-washing or checking the postage on a letter multiple times before putting it in the mail.[25] These are unusual groups when it comes to making decisions. If you have teenagers at home, you'd probably agree that their decision-making is, well, shall we say, unique among age groups. And while it's interesting to learn that women with obsessive-compulsive tendencies delay decisions more than their male counterparts, this cannot speak to the general population.

So the notion that women can't make up their minds is misleading. It's true that some women find it harder than others to make up their minds, but some men are just as stymied. The evidence indicates that most of the time, a healthy adult man struggles over a decision just as often as a healthy adult woman.

The Appalachian Trail with Shin Splints and a Stomach Bug

If women are just as decisive as men, why do so many people still see men as more decisive? Wouldn't we all notice if women leaned toward action just as often as men? There could be many reasons why this notion of indecisive women persists, but I'll highlight two that we all should understand.

Jennifer Pharr Davis wanted to hike the Appalachian Trail faster than anyone had ever hiked it. She'd already completed the 2,181-mile trail twice, and on her second time through, she'd set a new record for women, hiking an average of thirty-eight miles a day for fifty-seven and a half consecutive days.[26] For many of us, that would be enough. We would tend to our blisters, and, when we started itching for a challenge, we'd hike something new.

But Jennifer wanted to try it again, and her goal this time was to beat the men's record. As she recalls in her memoir, where most of this story appears, she wanted to get from Maine all the way to Georgia averaging forty-seven miles a day. She wanted to do it hiking, not running. She talked about it with her husband, Brew, who, as her support team, would have to spend two and a half months on the road with her while she hiked the trail. Each night, he would drive ahead to their next rendezvous point and be ready with dry clothes, a warm dinner, ample Gatorade, and a set-up tent. Brew was a schoolteacher and had the summer off, but he'd still be spending that time supporting her world-record attempt. He said if the decision were up to him, he wouldn't do it because it was so much pressure on both of them, but he would support whatever choice she made. She decided she needed to try.

The first few days of the hike were fine. She made great progress, averaging more than forty-eight miles a day, exactly as she'd hoped. But then her legs began to hurt with every step. It was only day three and she had shin splints. A bad rainstorm blew in as she hiked Mount Washington, and the dropping temperatures and wind on Franconia Ridge in New Hampshire gave her hypothermia. She kept going. By day seven, she'd caught some kind of stomach flu and couldn't hike more than an hour without stopping to go to the bathroom. The six hours of sleep she managed each night weren't allowing her body to heal.

On day nine, she decided she was done. She'd started out that morning strong, but she kept getting sick on the trail. She sat down, too weak to continue. Her mileage had been dropping for days, and she wouldn't beat the overall record at this rate. The decision not to press on felt like sweet relief and although it took her hours longer than it should have, she finally made it to the trail rendezvous point where her husband was waiting. She told him, "My body can't do this anymore. My shins hurt and I'm still sick to my stomach. I don't think we can get the record. I just want to stop."

She had no doubt that Brew was going to tell her that it was okay; she was sure that he would hold her for a while and then walk her gently down to their car and drive her to a hotel where she could finally take a hot shower.

But he just looked back at her and said, "You can't quit. I'm not going to let you." He kept talking, but she missed what he said next. She was so shocked she couldn't listen. When she finally tuned in again, she heard him saying, "You've given too much to the trail to quit now. You owe it to yourself to keep going. And, a little bit, you owe it to me. If you want to quit, if you *really* want to quit, that's fine. You can make that decision tomorrow, or two days from now when your stomach settles. But right now, you need to eat and take some medicine, and you need to keep hiking." If she made the wrong decision, he emphasized, she'd regret it for the rest of her life.

His blunt, unyielding response made her angry, but she ate dinner, took some medicine, and slept. She weighed what he'd said. She decided it made sense to wait and see if the medicine worked. Jennifer realized that she owed it to the team—to herself and her partner—to try a little harder, so she reframed her goals and decided to do her best and discover whatever that might be, even if she couldn't beat the world record. The next morning, she started hiking before sunrise, and after twelve miles, the medicine finally kicked in. Two days later, she couldn't imagine that she had ever wanted to quit. As we'll see, she's glad he insisted.

What goes through your head when you hear Jennifer's story? Some people will be inspired by the fact that she changed her mind and decided to persevere. Some will think she has an incredible team mentality, that she must deeply trust and respect her husband's judgment to even entertain the idea of getting back on the trail after she had made her peace with her decision to quit. I have no doubt that Jennifer loves her husband and that he loves her back. In fact, she titled her memoir *Called Again: A Story of Love and Triumph,* because for her, hiking sixteen hours a day for nearly twenty-two hundred miles says as much about her love for her husband as it does about her will and endurance.

But not everyone will be impressed with Jennifer's decision. Some people will read her story and ask angrily, "Where's her backbone? She was sick and in immense pain and should have listened to her body. This was her dream and stopping should have been her decision." Someone who attempts the Appalachian trail in its entirety on three separate occasions is anything but weak-willed. And yet that's the judgment trap many women find themselves in. Critics interpret an openness to input as indecisiveness; they assume that when a woman doesn't decide on her own, it's because she can't.

If She's Collaborating, She's Not Being Decisive

Before we reach the happy ending of Jennifer's story, let's consider these conflicting reactions, one laudatory and one critical, that observers often have when they watch women take a team mentality: Some celebrate it, while others call it weak. These mixed reactions happen in the workplace as well. We heard it in Nina's story, where rumors circulated that female managers made their decisions based on the last people who talked to them. We read it in industry analysis too. In a report by McKinsey and Company, a European top executive said: "Women tend to be more participative in their decision-making compared to men, which is sometimes perceived as a lack of ability to make decisions."[27]

This, I believe, is one of the primary reasons that women are seen as less decisive: people assume that being decisive and being collaborative don't mix. Take the day Jennifer had to decide whether to quit or keep going. She weighed her husband's input as seriously as, if not more seriously than, she weighed her own. And notice that even though she decided after several days of misery that she needed to stop hiking, she didn't announce it as a resolute *Scratch the original plan. Here's the new plan.* She said, "I don't think we can get the record. I just want to stop." Whether she meant to or not, she left room for his input.

Research shows that leaving room for input is more common among women, especially when female leaders are compared to male leaders. Whether they are studied in the workplace or in a controlled laboratory experiment, women in leadership roles tend to take a more democratic approach to their choices and consult more people in the deliberation process than their male counterparts. Presumably a woman's life partner has more influence than most others, but closeness isn't a prerequisite. A woman randomly assigned to lead a group of strangers consults those strangers more often than a man in the same role.[28] Female mayors are more likely to involve citizens in budget decisions, and female managers allocate more time in meetings for employee feedback.[29]

Does this observation that women collaborate more often than men mean that every man will dictate his decisions while every woman will take a poll first? No. There are plenty of egalitarian men who seek group buy-in, and there are plenty of women who assign tasks without asking a single soul in the room for input. And within an organization, a culture or a set of policies could ensure that decision-makers consider multiple perspectives. But in the absence of those practices, in most organizations, we expect to find women taking a more egalitarian approach to leadership decisions than men.

You're hardly shocked. Most of us expect women to be more collaborative than men. We expect women to care what other people think, to be more focused on the team, and to give up their strong positions to accommodate others.[30] When researchers asked people, "Who do you think would be better at working out a compromise, a male political leader or a female one?" those respondents who felt there was a difference picked the female politicians four times more often than the males.[31] It fits the mother-duck model. She wants to go this way, but everyone else wants to go that way, so rather than venture out on her own, she strikes a compromise. This is the bind many women find themselves in — be democratic and be seen as indecisive and weak-willed, or be authoritative and be seen as decisive but selfish.

Women might be more democratic in the workplace because that's the only way they can lead. When a woman tells coworkers or subordinates what to do, they don't like it, and they usually don't like her. Studies have found that people penalize autocratic women, rating them lower than men on performance reviews, even when men give the same marching orders.[32] Autocratic women also receive greater blame when an organization fails. A 2014 experiment found that when a woman dictated what should be done and a division subsequently became less productive, observers blamed the woman.[33] "She was incompetent," they charged, or "She didn't work hard enough." But when a man dictated a key decision and a division then produced the same poor numbers, observers rationalized that it was a combination of his poor leadership and factors out of his control. Men weren't entirely off the hook, but critics were more willing to blame the circumstances.

"The stock market was down," they said, or "Look at all that his team had on their plate." What about women who didn't dictate the strategy, who took a collaborative and democratic approach and involved others in the decision-making process? These women were seen as competent even if the group's performance declined. Only when a woman was collaborative were critics willing to point a finger elsewhere. "She listened to others and she asked her team, so it must have been bad luck or bad timing," people said.

Why are we more accepting when a man tells people what to do than when a woman does? Alice Eagly, a psychologist at Northwestern University, and Steven Karau, a management professor at Southern Illinois University, suggest it's due to what they call role congruity theory.[34] The idea is that we like people more when they behave in ways that fit society's typical roles for them, and we like them less when they don't. We expect men to behave like managers — setting priorities, saying no, and making those difficult decisions that some (perhaps most) people won't be happy about. But the typical expectation for women — that they're helpful, nurturing, and sympathetic — doesn't fit the usual management role. We might like her as a manager, but only if she shows she listens to us.

The message is that a male leader can be autocratic and decisive and still survive a failure, but a female leader can't. She has to be collaborative if she wants to protect herself. At the same time, she won't be considered a good leader if she's not decisive.

Does Being Collaborative Lead to Better Decisions?

So it might weaken a woman's reputation as a decision-maker if she's open to input and influence, but does it lead to better decisions? In Jennifer's case, it did. Once she felt healthier, she was able to pick up her pace and average almost fifty miles a day for the rest of the trip. It's hard for me to imagine walking fifty miles in a single day, let alone hiking that distance for forty-six consecutive days. Even though she'd been terribly sick, she completed the trail in record time, beating the record previously set by a man not by five minutes or even five hours but by over twenty-six hours.

Does a shared approach to decision-making lead to better choices in the workplace? That's harder to evaluate. Scientists find that leaders asking others for input results in better decisions in some situations. For instance, mixed teams — teams with members from different backgrounds or departments — benefit from collaborative decision-making. If you're going into a meeting that brings together people with different areas of expertise — perhaps a graphic designer, a chef, and a publicist — you're going to walk out of that meeting with a more innovative plan if you give everyone an opportunity to weigh in than if you dictate the decision you want.[35] Team members are more likely to offer their specialized knowledge when they feel you'll listen.[36] Your chef might be able to warn you that pork belly is so yesterday, but if you take an authoritative stand, your chef might say nothing.

At least one study suggests that women bring something special to the collaborative decision-making process: the willingness to seek advice. Ask someone if he or she knows a man who won't ask for directions, and you're bound to hear a story about a father, brother, or spouse who drove miles out of his way while a woman in the car kept suggesting he stop and ask for help.[37] Maurice Levi, a University of British Columbia business professor, wondered whether men and women differed in their help-seeking in professional contexts. Levi led a team that looked at one of the biggest decisions a company can make — whether to allow itself to be acquired by another company. You might not expect gender to play a role in this kind of decision. After all, asking for advice on a merger and acquisition deal is very different than asking for directions to the nearest gas station from your minivan. Levi's team looked at companies that had been targeted for acquisition and asked whether they relied solely on the internal judgment of their executives and advisory board or whether they sought external help.[38] Levi's team found that almost every company hired a financial adviser when it received a bid. With so much at stake, companies wisely wanted advice. But companies with more women at the top sought advice from first-rate sources. Target companies with women on their advisory boards were more likely to hire one of the top twenty nationally ranked merger and acquisition advisers. For every 10 percent of the board made up of women, the chance that the company sought a top-ranked adviser went up 7.6 percent.

What does it mean that companies with women went with the top-notch advisers? One possible explanation is that companies with women on their boards put a higher value on outside advice. The highest-ranked financial advisers in the country, companies such as Goldman Sachs and Morgan Stanley, probably charge higher fees than lower-ranked firms, so companies with more women might have budgeted more money to spend on outside advice. Another possible interpretation of all-male boards hiring second-tier advisers is that these decision-making bodies were extremely confident in their own ability to assess the offer but still felt they had to check a box. They dutifully hired consultants, but because it was a pro forma exercise, it wasn't worth investing in the top tier. And if you're thinking that people are more likely to follow advice when they pay a lot for it, you're right.[39]

Eugenia, a chief marketing officer at a data visualization company where many employees are in their twenties and early thirties, saw additional reasons that it was essential to be collaborative as a leader. "In today's work force, being able to make people part of a decision, or at the very least assure them they've been heard, is crucial. Especially with the millennials, they need to be brought in." She also believes that rather than slowing down a process, collaboration speeds it up. "When a woman is being collaborative, what she's really trying to do is think three steps past the decision. A decision isn't just a decision. It's a decision that has to be implemented. So if I spend a week gathering input from people, guess what? At the end of that week, that decision is going to go down a lot better and a lot faster. There's no 'let's revisit this' or dragging of feet when it's time to implement," Eugenia said.

Is it always better to ask for input before you act? Of course not. If you're the lead surgeon in an operating room, you don't want to ask your team what to do when your patient's vital signs begin to drop. Not only would it waste precious time, but if you're the most experienced and knowledgeable specialist in the room, a group vote on the best course of action might mean that the patient suffers or dies while you negotiate. In operating rooms, the best decisions are made when the person in charge decides autocratically.[40] We can expect the same is true in other professions where there are clearly defined optimal choices and there isn't the luxury of time to confer.

What Price Do Women Pay for Being Collaborative?

So is the end of the story that women are collaborative and that makes the world a better place for everyone? Yes and no. If we're looking at what's best for organizations that want to capitalize on different experts, it's wonderful that women are collaborative. But if we're looking at what's best for women, being collaborative has serious drawbacks. From a perception standpoint, it feeds into the indecisive label. People are primed to see women as indecisive, so even when a woman hesitates for the best possible reasons, when she's seeking premier advice or when she's getting employee buy-in, her reputation suffers. She pays a personal price for using a decision-making process that leads to a better outcome.

There is a high expectation that women will collaborate, and that expectation places an added burden on women. When Loretta Lynch, the top federal prosecutor in Brooklyn, New York, was nominated for attorney general of the United States, a prominent defense attorney, Gerald L. Shargel, praised Lynch for her availability, saying, "Any time I had an issue with a case, and thought it was appropriate to knock on her door, she was welcoming."[41] Most of us would take that as a compliment. But what do such women have to sacrifice? How often did Lynch have to set aside whatever she was working on to ensure she stayed collegial? We'd never expect anyone to make that observation about a powerful male decision-maker. No one says, "President Obama's door is always open," or "Whenever I have an issue with Facebook, Mark Zuckerberg has the time." Men are not usually judged by their availability. If someone did say a male leader was always welcoming and willing to talk at the drop of a hat, we might find it refreshing, but we might also question whether he got any real work done.

I heard women describe this high bar for collaboration in my interviews as well. Kat, a former CEO, was making a beeline for the restroom one day, and an employee stopped her and asked a question. Kat gave a short, direct answer — "Yes, do it" — and she later got an e-mail from this employee asking if Kat was upset, if something was wrong; why didn't Kat want to discuss the issue? One possible explanation for the concern is that for a CEO, male or female, every nuance is noticed. But an added expectation is that a woman will want to stop and discuss things, never mind what else she's doing or how urgent it might be.

The expectation that women will collaborate opens a complex issue and a trap for many women leaders. We've all heard what happens to women who fail to meet this expectation and aren't considered inclusive enough. Women who make executive decisions and tell people what to do (instead of asking them what they'd like to do) win points for being decisive but lose far more points for likability. The press (and plenty of others) call these women bossy, pushy, and opinionated, and the adjectives quickly get nastier from there. This is what Joan C. Williams and Rachel Dempsey call the Tightrope Bias in their book *What Works for Women at Work*. Women walk a tightrope between masculinity and femininity; society criticizes women leaders for being bossy if they act like traditional men but tags them as indecisive and poor leaders if they act like traditional women.

Strategies for Decisive and Collaborative Can Mix

How can a woman show that she's both decisive and collaborative, that she doesn't have to sacrifice one for the other? Framing how a decision will be made can help. If you're a woman in charge of a project, tell people right from the start how you will proceed. Strive for active transparency. Perhaps you say that you'll carefully consider different sides of the argument, that you want people to bring evidence to the table, that you want to hear the specialized knowledge certain groups might have that you don't, but that, ultimately, you'll decide what you think is best for the team.[42] Clarify how long you're taking input so that people have a chance to influence the decision but won't be stopping by with ideas down to the last minute.

Erica, a software engineer, struck this balance by proposing a solution and then asking for concerns. In a recent meeting, she said: "You've all read the report that I sent around. We can review the research, but I need to move with this decision. Management, of course, wants a solution yesterday. So, before we go into analysis, are there any objections to our doing this? Does anyone see a reason that it's not a good fit for us?" She said that by framing the discussion this way, she communicated that the focus was on action, on moving forward, but that she also wanted to hear concerns if they were worthwhile.

It may help to reframe how you're approaching an important decision. For example, try thinking, *I'm a contextual decision-maker. I want more information because the context matters.* Economists Rachel Croson at the University of Texas at Arlington and Uri Gneezy at the University of California, San Diego, made this observation when they compared men's and women's choices in a variety of settings.[43] They found women were much more sensitive to changes in context; women responded to how many people would be affected by a risky decision, whether they could make the decision anonymously or it had to be public, and so on. When I've told women, "Perhaps you're being a contextual decision-maker," they've found this empowering. Some women are so used to the message "You can't make up your mind" that it helps to have a counter-message: "Actually, I'm considering the context."

Another strategy involves being selective about when you ask for input. If you think you might be pigeonholed as the collaborative type and therefore seen as a bit too reliant on the group, ask for input on some decisions and openly make others on your own.[44] Be selective as well about who actually needs to be brought in on the decision-making process. In Nina's situation, where people believed that female managers were most strongly influenced by the last people they'd spoken to, perhaps the managers hadn't successfully communicated whose input was still needed and whose views had already been taken into consideration. Of course, those managers might have said quite plainly, "Thanks, but I've got all the input I need on this decision," but the employees lined up anyway. That's one of the problems with the motherduck view of women leaders. If you feel like a straggler duckling, you can't possibly believe that she's moving ahead without you. There must be some mistake.

So the first reason that women are seen as less decisive is that they often collaborate. Once dubbed collaborative, considerate, or inclusive, they are also dubbed indecisive. But does the fear of being labeled indecisive actually influence how women decide? Do women make decisions any differently when this label is hanging over their heads than when it isn't? This brings us to the second reason why I suspect women are seen as less decisive: the stereotype really does get maddeningly in their way.

What Can Happen If the Job Description Is All About Gender

Diane Bergeron, now a business professor at Case Western Reserve University, wondered whether women actually made decisions any differently when they were reminded of how people see men and women. Perhaps notions of decisive men and indecisive women bounce off the women, leaving no mark. Perhaps they stick and interfere with women's decisions. Or perhaps these labels make women double-down and do amazing work to prove everyone wrong. Which is it usually?[45]

Bergeron took a group of aspiring professionals — male and female graduate students who were training to become human resource managers — and gave them twenty-four management decisions they might have to consider on the job, decisions such as whether or not to approve a questionable reimbursement request, which employees should be promoted, and how to proceed with a sexual-harassment complaint. Then Bergeron started the timer, and each student tried to make as many good decisions as possible in forty-five minutes.

So where do the labels and expectations come in? This is where the study becomes brilliant. Bergeron didn't come right out and say, "Women choose poorly." People rarely say that so plainly. Instead, she reminded the budding HR managers about the stereotypes in a way that might be realistic for a twenty-first-century American workplace, where no one is allowed to blatantly discriminate against women. Bergeron played with the job description. For half the men and women, she said the previous manager in this job was a woman and described the ideal candidate with stereotypically feminine adjectives — nurturing, understanding, intuitive, and aware of other people's feelings. For the other half, she said the previous manager was a man and suggested the ideal candidate would possess stereotypically masculine attributes — he'd be aggressive, self-confident, achievement-oriented, and, of course, decisive.[46] She was not saying that this was how men and women truly were but that these were the stereotypes.

Imagine you're participating in this study. You're looking to prove yourself, you have a stack of decisions to make, and you've been told that showing good judgment in this job equates to being a caricature of a feminine woman (or a caricature of a masculine man, depending on the group you're in). Would that job description matter? Would the ideal you have in your head affect what you did? We can all probably agree it shouldn't — whether or not an employee is eligible for a promotion should depend on that person's job performance, not on who you think you should be when you grow up.

But the description of the ideal job candidate did matter. First, the good news: The women who had been encouraged to think of a feminine role model showed strong, focused decision-making. They worked their way through a greater number of decisions than the women who had a masculine role model. How did they compare to the men? The women with a feminine job description made just as many decisions as the men, regardless of who the men were trying to emulate.

Now for the bad news: The women in this study who had been led to believe that the ideal person for the job was a man, and a highly masculine one at that, moved through the decisions more slowly than any of the other groups. They showed hesitation and indecisiveness. On average, these women made 15 percent fewer decisions than the women with a feminine role model, and they made 9.5 percent fewer decisions than men in any group.

Making a lot of decisions is one thing, but quality isn't the same as quantity. Were quality decisions being made? When women in this study felt pressured to fill a man's shoes, they were especially likely to make mistakes on complex decisions. Consider this scenario: One message specified that anyone cited in a sexual-harassment complaint would be under investigation, and another message, nine or ten down the stack, indicated that if an employee was under investigation, he or she would be ineligible for a promotion that year. Then came the decision they had to make: Could Isaac, an employee who, according to a third memo, had been charged with sexual harassment, be considered for promotion? No—approving Isaac as a candidate for promotion would be a poor decision. It seems simple enough when I put those facts side by side, but if you were in the study and that timer was running, you wouldn't see the information laid out this way. You had this and almost two dozen other facts to hold in mind simultaneously. So which subjects said that Isaac was eligible for promotion? That mistake was made most often by women who had been nudged to believe men did this job best. These weren't errors common to all women— just the ones given a masculine job description. Women with a feminine job description made just as many, if not slightly more, excellent decisions in these complex scenarios than the men in either group.

What happened if you were a guy trying to fill a successful woman's shoes? Did those men face the same problem with indecisiveness and poor judgment? Not at all. In fact, both men and women who were trying to behave like a successful feminine candidate showed better performance than those who were trying to behave like a masculine one. What was so special about the feminine job description? The subjects had been told she was nurturing, understanding, and empathic, and as we learned in the previous chapter, when people try to imagine what others need, they often make higher-quality decisions (and one could see how that would be especially true in HR).

But let's go back to the women who struggled with their decisions, who became indecisive when they were told they were doing a man's job. Why would these women perform so poorly? No one made any derogatory comments. And certainly no one said, "Women suck at this." Are human resource personnel just easily intimidated by job descriptions? No; this is an example of a much more common problem, one that cuts across professions. It's an example of something called stereotype threat, and if you're a woman or a minority, you've probably faced it.

Stereotype Threat and the Burden to Prove Them Wrong

Stereotype threat has become one of the most important and highly studied concepts in social psychology in the past two decades, yet a surprising number of people still don't know what it is or how it affects them.[47] They don't realize it has the power to explain why they suddenly find it hard to make a choice around certain people or why they grow quiet in certain meetings. They don't know that stereotype threat can make them choke under pressure and render them momentarily unable to decide anything.

Stereotype threat is the anxiety that you'll live up to someone else's negative expectations of your group.[48] Think for a moment about a group to which you belong. You might be young or old, heavyset or rail-thin, straight or gay, an advocate of mass transit or an owner of three cars—the list goes on. Next, think of a negative stereotype about your group, something that may be wildly inaccurate but that you know other people believe is true. Chances are you can generate at least one if not several negative stereotypes; all groups have a few, and some groups are burdened with many.

Whatever that derogatory and unwanted belief about your group might be, stereotype threat is the fear that you're about to do something that shows it is true. It's the anxiety that some action you're about to take will pigeonhole you, make you the disappointment that someone else has already pegged you, and everyone like you, to be. So falling into expected boxes means you're failing to properly represent yourself, but it also means you're failing your community. And because part of you is chewing on this possibility, you don't live up to your abilities.

Here's my own experience: I enjoy driving. There are a few males in my family who have commented, on multiple occasions, that women can't drive well. Most of the time, I don't think about this stereotype when I'm driving because this belief is preposterous. Statistically, men break more traffic laws than women and have a greater number of fatal accidents.[49] However, if I'm behind the wheel and one of these lovely gents happens to be a passenger, I make different driving decisions than I normally would, even if he doesn't mumble a word about women. I am hypervigilant at every turn. I'll choose an easier route rather than the shortest one, I'll take fewer risks when I'm merging onto highways, and there isn't enough soothing jasmine tea in the world to make me attempt a parallel-parking job. Relieved and exhausted when I finally turn off the car, I usually hand the keys to someone else for the return trip. I love to drive, but I don't want to feel as though my every decision is being judged and scrutinized. The pressure of perfection is too high. I feel bound to fail and fulfill their low expectations. Which makes me a worse driver. That's stereotype threat.

If you've heard of stereotype threat, you might remember that it's often used to describe a problem that African American students run into on achievement tests. The notion of stereotype threat originated in 1995, when two psychology researchers at Stanford University, Claude Steele and Joshua Aronson, wanted to understand the achievement gap between African American students and white students in the United States.[50] Disheartening but true, African American students typically don't do as well as white students on standardized tests in the United States; from grade school all the way up through graduate school, the average standardized test scores of African American students lag behind those of their white peers.[51]

Steele and Aronson wondered if they could reduce that difference by making a tiny change in the instructions on a test. They invited sophomores at Stanford to sit down and answer thirty questions from the advanced section of the GRE, the test that college seniors, not sophomores, take to get into graduate school. By picking the hardest questions on the test, they knew even these extremely capable students would struggle. That part worked — the students were challenged and on average answered less than one-third of the questions correctly.

Here's the innovative part: Steele and Aronson played with how the students saw the test. They told half the students the test measured their intellectual abilities, and their score would reveal their true strengths and weaknesses. Imagine yourself in this situation. You're a successful college student at a prestigious private school. The person passing out the test explains that this test shows how intelligent you are and will reveal where you have potential and where you're not as capable. Does this test matter to you? Chances are it does, whether you admit it or not.

They gave the remaining students a different story. Steele and Aronson told the second group that the research team was studying how people solved problems, explicitly adding, "We're not going to evaluate your ability." That's all they changed. At first glance, those instructions don't seem that different. You sit down to take the test and you're told either "We're measuring how intelligent you are" or "We're studying how people solve problems; we won't be measuring your intelligence."

For the white students, the explanation didn't make a difference; they performed basically the same regardless of what the researchers said they were testing.

But for African American students, the difference was dramatic. African American students who heard "We're measuring how intelligent you are" performed worse than white students, earning scores that were 25 percent lower. Something got in their way. What about the African American students who heard "We're studying how people solve problems; we won't be evaluating your intelligence"? They earned scores that were just as high as the white students'. The achievement gap between the races had, just like that, disappeared.

Countless studies at other universities have replicated these findings.[52] Although there's some debate over the nuances, the consensus is this: When African American students take a test of intelligence, anxiety that they might fit an awful stereotype gets in the way of their performance. In the United States, there's a pernicious and far too prevalent belief that African Americans aren't as book-smart as whites, so when African American students sit down to an intelligence test, they have a lot on their minds.[53] They're anxious that the derogatory image might apply to them, and instead of concentrating on the test, they use some of their mental resources to manage that anxiety. When they become frustrated and can't answer a question, they don't just think it must be an unusually hard question, shrug it off, and move on. They think, *Damn it, I don't have what it takes,* or *No, stay calm, you're smarter than that.* They're not anxious about the test alone, they're anxious about the label. Even when African American test-takers aren't worried about their own intelligence, they still feel pressured to prove how wrong the label is for African Americans in general.

What about the African American students who thought the researchers were just studying how people solved problems? There's no stereotype about solving problems, no looming label. Those students had the same experience and the same results as the white students — they had to concentrate only on the extremely challenging test.

The other well-known example of stereotype threat involves women taking math or science tests.[54] There's a long-standing, hard-to-shake myth in Western countries that males are naturally better at math than females. As early as age ten, children in the United States and much of Europe already "know" that boys are supposed to be stronger at math. Sadly, young girls often believe it's true, even if their own personal grades in math are outstanding.[55] By the time these girls become adults, the notion that women struggle more with math is so ingrained that it takes very little to ignite their performance anxiety. One study found that asking adult women just before a test "How will other people view you based on this difficult math test you're about to take?" will make their scores fall by an average of 59 percent. That means a woman who's worrying over how she'll be judged could drop her score from a 10 to a 4.

Wouldn't anyone be shaken by that annoying question? Evidently not. Ask white men to consider how others will evaluate them based on that same test score and they don't flinch.[56] They perform as though no one said a thing. Even men who aren't particularly good at math — men who generally avoid math courses in college — still cruise through the test unperturbed when you remind them that the world might judge them based on their scores. Do white men ever become stigma conscious on a math test, enough that it hurts their performance? Yes — when you explicitly tell them that Asian students tend to be phenomenal at math and outperform Caucasian students on quantitative tests.[57] Mention that stereotype before a test starts and even those white males who are strong math students will make mistakes.

It's a simple formula: Take individuals who normally perform well, remind them they're part of a disadvantaged group, and watch them falter. It's not fair. But it happens when we run smack into a stereotype.

For women, the anxiety that they'll live up to society's low expectations becomes a formidable distraction to manage.

And this phenomenon, Bergeron says, explains what happened to women in the human resources study. We can picture what was going through their minds. When women heard the message "The ideal candidate for this job is an authoritative guy" and then found themselves struggling with hard decisions under time pressure, they began to get anxious. *This is really difficult. Maybe I'm not cut out for this after all.* Of course, anyone can be distracted by intrusive doubts. But women who'd heard that the ideal person was a decisive do-this-now masculine type were especially susceptible because the message echoed a societal belief that women are slower to decide, that they struggle with their choices. Distracted by these unwelcome anxieties, they took longer to make decisions, probably doubling back to reread some of the messages multiple times. In the end, they made fewer decisions and poorer ones, confirming the stereotype of indecisive women.

So this becomes the second reason that society sees women as less decisive. If the people around a woman communicate that women take too long to choose, that what's needed right now is a decisive, forceful man at the helm, then there's a very good chance that when she faces a really difficult decision, those stereotypes will come back to taunt her. The negative expectations make her hesitate; they create bubbles of doubt that wouldn't be there in a more supportive or neutral environment. When people expect women to second-guess themselves, then, like the worst kind of magic, women will.

Is Stereotype Threat Happening to You?

It's easy to misunderstand stereotype threat. You might think, *Ah, it's a self-fulfilling prophecy,* but stereotype threat is more pernicious. With a self-fulfilling prophecy, you're the person who is predicting that you're going to perform poorly. With stereotype threat, you're battling someone else's beliefs.

Think this won't happen to you because you're committed to doing outstanding work? Sadly, stereotype threat is actually a bigger problem for those who take their work seriously than for those who phone it in.[58] Women who aren't deeply invested in what they're doing, who aren't trying to be great managers or stellar attorneys (or, as in my case, skilled drivers), aren't typically wondering what other people think of them in those roles. But if you're a woman who cares deeply about performing well, if your career is a central part of who you are, then you're more bothered when people hint that skilled, dedicated women are a rarity.

So is the solution for women to stop caring about their work? Hardly. But before we go into strategies for inoculating against stereotype threat, let's identify whether it's a problem for you.

Would you know if you were feeling anxious? Would you realize if you were affected by someone else's low expectations? After all, if you're trying to make a decision and other things in your life, such as your cell phone or your five-year-old or your plans for the upcoming three-day weekend, distract you, you might notice the problem and then either find a way to shut out the diversion or postpone making the decision until you can think without interruption.

But your own internal anxiety is very different from those external distractions. As human beings, we're notoriously bad at recognizing when we're anxious or when anxiety drives our choices. We misinterpret anxiety as a sign that we're angry, undervalued, overworked, or, oddest of all, attracted to someone. Researchers revealed that people can mistake anxiety for attraction in a classic study in which a young, pretty woman stopped men just after they'd finished walking across a high suspension bridge, the Capilano Bridge in Vancouver, British Columbia. The bridge makes people extremely nervous — it's wobbly and narrow; from end to end it's longer than a football field; and the sheer drop to the river below is more than two hundred and thirty feet. There's no question that each man felt anxious when he stepped back onto solid ground, and there he found a pretty woman who handed him a pencil and a questionnaire. After this attractive researcher asked each man to fill out the survey, she gave him her phone number and said he could call her later if he had any questions about it. The woman did the exact same thing at a nearby bridge that was much sturdier and didn't make anyone anxious. What happened? The men who had just crossed the nerve-racking bridge were much more likely to call the pretty researcher that night than the men who had crossed the sturdy bridge. They were anxious when they stepped off that wobbly bridge, but they didn't see their reactions that way. They misread intense anxiety as attraction worth acting on.[59]

As far as I know, a good-looking researcher didn't approach women who crossed the wobbly bridge, but other studies show women overlook or misinterpret their anxiety as well.[60] When women heard "Men do better on math tests" just as they sat down to take a math test, most said they didn't feel any added anxiety. They'd heard that men were supposed to be superior in math many times—but knowing something in the back of one's mind and having it cloud one's immediate judgment are two very different things. Despite the women's claims that they were calm, their bodies showed the pressure they were feeling—their hearts raced and their blood pressure went up.[61]

Even when people do realize they're anxious, they often can't pinpoint what made them feel that way. In Bergeron's study in which future HR managers were told that the best person for the job was a very masculine man (or a very feminine woman), women who felt anxious probably chalked it up to the pressure of the test, fearing that if they made a mistake, they'd be labeled below average at HR work. It's very unlikely that any of the women consciously thought, *I'm feeling anxious because I was told the best person for this job was a decisive male.* Yet we know it was precisely these women, the ones who were trying to fill a decisive man's shoes, whose decisions suffered. It was also these women who walked away from the experience saying they weren't interested in a career in HR anymore.

If we can't fully trust our bodies to tell us we're feeling the splinter of anxiety, how can we know? Look for cues around you. Stereotype threat is often triggered by cues in the environment, cues that become obvious once you know what to look for. The first cue in stereotype threat for women is the number of females in the room. If a woman is far outnumbered by men, she is more likely to become preoccupied with being the "only one."[62] That preoccupation chips away at her problem-solving skills. One study looked at what happened when three people sat in a room taking a challenging math test. Women who took the test in a room with two other women scored an average of 70 percent, whereas women who took the same test in a room with two men averaged only 58 percent.[63] In most American schools, that's the difference between a passing and a failing grade.

So for a woman, what's the magic number of females in a group? Unfortunately, there isn't one. In a small group, having just one other woman might be enough to reduce the anxiety and the pressure. When Justice Ruth Bader Ginsburg arrived at the Supreme Court, Justice Sandra Day O'Connor found that her working world changed.[64] O'Connor told Nina Totenberg in an NPR interview that when she was the only female Supreme Court justice, the media experience was "asphyxiating." "Everywhere that Sandra went," O'Connor said about herself, "the press was sure to go." After every newspaper write-up of a Supreme Court decision, "there would be a little add-on: What did Justice O'Connor do in the case?" Being singled out made O'Connor hyperaware of her every decision, and that went on for twelve years.

But when Justice Ginsburg joined the court, in 1993, the routine scrutiny stopped. O'Connor was no longer an anomaly. "Oh, it was just night and day," O'Connor explained. "The minute Justice Ginsburg arrived, the media pressure was off . . . We just became two of the nine justices."

Of course, for a woman in a room of a hundred people, one other woman won't be enough. What's needed is what's termed *critical mass*, but there is no prescribed number.[65]

The proportion of women to men isn't the only trigger. (In fact, some women aren't affected by the ratio of men to women, at least not all of the time.) What else triggers stereotype threat? To help women assess their environment and how they respond to it, here's a short list of questions.[66] (I've designed these questions for women and the stereotypes about them, so if you're a man reading this, even if you're an extremely empathetic man, you might want to hand this to a female friend if you're curious how she experiences your work environment.) Answer each question with True or False, picking whichever answer applies more often than not.*

___ 1. Some of my male colleagues believe that women are not as committed to their careers as men.

___ 2. When I think about my career progression, I often compare myself to the men in my organization.

___ 3. Sometimes I worry that my behavior will cause my male colleagues to think that stereotypes about women apply to me.

___ 4. I feel I am continuously switching between my feminine self and my work self.

---

* Please note that this is an informal list of questions, not a validated test of stereotype threat. I developed the questions based on environmental cues that researchers have identified as indicators of stereotype threat.

__ 5. If I make a mistake at work, I'm sometimes concerned that my male colleagues will think I'm not cut out for this job because I'm a woman.

__ 6. Some of my colleagues feel that women have less ability than men in this job.

__ 7. I work in an environment that prizes stereotypically masculine qualities, such as being decisive, aggressive, self-reliant, self-confident, and achievement-oriented.

Now add up the number of True answers. The higher the score, the more likely the woman is to experience stereotype threat at work.

How high does the score have to be for a woman to be concerned? I wish I could answer that, but it's going to depend on the individual and how blatant the cues are. One study found that for college-age women, it took only two cues to trigger stereotype threat: being the only woman on a team of four people and having someone blatantly state that women didn't perform as well as men.[67] What if there's just a single cue? That wasn't a problem, according to the study. Simply being the solo woman was fine as long as no one said women couldn't achieve; likewise, hearing someone strongly state that women couldn't achieve didn't bother women as long as women made up 50 percent or more of the team. But put those two cues together, and women became anxious and their performance declined. Of course, these were students in their early twenties, and experienced career women might not be as easily ruffled. But these findings do suggest that as the cues accumulate, so does the impact.

If you're a woman, perhaps you're thinking that you don't run into this problem with most choices you make each day. Indeed, stereotype threat isn't a problem for easy decisions. When you're deciding what time to go to lunch or whether it's better to send an e-mail or a text to a friend, you're not likely to feel anxious that you'll be viewed in a particular way.

Stereotype threat comes to bear with hard decisions. It happens for choices that require one's full attention. If a woman is trying to decide how to cut the budget by 40 percent or how to handle a high-profile project that's becoming an embarrassment to her organization, chances are she will feel frustrated. Anyone would. But when you're part of a stigmatized group, you're more likely to interpret frustration as a sign that perhaps other people are right—maybe you're really not cut out for this. It's the high-stakes decisions that trigger a cascade of doubt.

Why Ignoring That Splinter Doesn't Erase the Pain

How does hearing "Men are more decisive" or "Asian Americans are better at math" or "White Americans receive higher test scores" make anyone unable to think clearly? Can't a talented person just push those thoughts aside and get back to work?[68] Yes and no.

Let's imagine that you're a member of a stigmatized group and you've just been reminded of a negative stereotype about that group. First, the bad news: When you're making a hard decision, you can't wave away these intruding thoughts the way you might ignore your cell phone. Part of it has to do with the nature of the distraction. This is a threat to your identity, to how you see yourself. It creates added anxiety and apprehension that you're going to be unfairly evaluated. You may not even consciously realize in the moment that people are propping up one group or putting down another, but the cues around you have put you on alert. You're now monitoring your environment more than you were before that comment was made.

The other part has to do with what a hard decision requires from the decision-maker. Many questions that face a leader—for instance,

"How can we spend less on coffee?" — require minimal mental effort to resolve. But a hard question, such as "How do we adapt to minimum wage increases?" or "How do we reduce infections in the hospital?," requires the leader to evaluate multiple pieces of information at once. Most of the information would be on paper or on the computer, but at some point, he or she would need to mull over all the competing factors.

Complex decisions like these tax something called working memory. Working memory is mental space where you maintain and manipulate all the facts, insights, and reactions you're considering at any given moment.[69] Try spelling the capital of Germany backward. Or take a moment and calculate 132 minus 67. For both of these, you're using working memory (or a piece of paper). It's called *working* memory because it's your private, portable workspace, an area for everything from mundane arithmetic to brilliant decision-making. It's a calculator and a whiteboard and a personal tape recorder all rolled into one. Despite all that flexibility, working memory is not infinite. Most of us can hold only about seven items in working memory at once, and as soon as you try to cram in anything beyond your personal limit, something else falls off.

A complex decision uses up your entire working memory. In fact, for many complex decisions, you max out at three or four items because each fact or insight is so big, much more complex than the letters of the word *Berlin*. Let's say you're a junior female administrator working at a hospital that's just made headlines for having one of the highest hospital-acquired infection rates in the state, and you're on the task force that needs to decide how to reduce those infections. You'll be considering which types of infections are common, which do the most harm, and which are easiest to control. You'll be thinking about how to balance immediate, easy fixes that reassure the public with slower, structural overhauls that will have a greater long-term impact. If you're politically minded, and you should be in this case, you'll be thinking about which powerful people will resist change and who will lead the charge. Your working memory is running at full capacity. With each new issue, one of the previous issues you were thinking about gets set aside, so you can't think simultaneously about the six most commonly transmitted infections, the four doctors who'll refuse to change, and the two newspaper reports saying that elevator buttons and grab rails weren't being cleaned.

Feeling overwhelmed? Okay, now let's add two more pieces of information: the ratio of men to women on the task force is seven to one, and in the first meeting, one of the elderly men remarked offhandedly, "We can't leave it up to any of the nurses to make decisions for their staff. Most of them can't even decide on a parking space." He didn't say the word *women*, but it was clear as day. You might get angry, you might roll your eyes, but the gauntlet has been thrown down.

Now, when the discussion of infection containment begins and you can't think of a viable solution, you wonder, just for a moment, if you're having problems not because it's a complex issue but because you're a woman. Unwelcome thoughts jump in, taking up one (or more) of your precious working-memory spots. You come up with a plan, but then you wonder, *Will they think this is stupid?* Suddenly, you forget the beginning of the idea you just had.[70] When you can't offer anything that hasn't already been shot down, you think, *When will I get another chance like this to prove myself?* The unhelpful self-talk continues, taking up mental workspace you need for the difficult problem you came here to solve. You say less in that meeting, not because you don't have ideas but because there's so much you're juggling internally.

Let's return to our original question: Can't you chase out those intrusive thoughts each time they come up? You can, but suppressing a thought also occupies working memory. You actually use up one of your few working memory spots every time you say to yourself, *Ignore the earlier comment; that person is an idiot.* Your mind is literally preoccupied and you find it harder to keep everything in sight.[71] Meanwhile, the people around you who aren't feeling that threat have all of their working memory available, which means they have both a head start and an ongoing advantage as decision-makers and problem-solvers. For women, that can mean the men around them seem to be gliding ahead with much less effort, which just fuels their anxiety.

Stereotype threat creates a difference between men and women that wouldn't otherwise exist. Make women feel that they have something to worry about, something to prove, and that preoccupation pulls their attention away from the decision. Take away that threat, and women decide just as quickly and effectively as men.

Anti-Inflammatories to Minimize the Threat

Now for the good news: You can protect yourself from stereotype threat. Geoffrey Cohen, a psychology professor at the Stanford Graduate School of Business, has said that strategies for reducing stereotype threat are like anti-inflammatories: you prevent the reaction, the ruminations and anxieties and thoughts and feelings that get in the way of good problem-solving.[72] A hard decision will still be hard, but you can prevent it from becoming any harder than it has to be.

Let's start with a question that might be on your mind: In the case of stereotype threat, is ignorance bliss? Is it better to know about these threatening cues or does that simply make things worse? After all, women learning about stereotype threat could exacerbate the problem because now they see their work environment differently. A woman might sit in a meeting wondering, *Do they think I'm indecisive? If I tell my boss I need a week to get back to him on a decision, will he think that I'm being a typical woman who has to ask around first?* Or she might not want to hear about stereotype threat, fearing that it's going to be just another thing she has to worry about, another obstacle she can't control.

But I promise, I wouldn't have gone into this much detail on stereotype threat if learning about it would get in one's way. Research shows that knowing is half the battle, and once people know about stereotype threat, they're less affected by negative cues in their environment.[73] In many studies, women who had just learned about stereotype threat performed as though the negative cues around them had vanished. How does knowing help? Being able to label your anxiety allows you to acknowledge that it's anxiety and keep that anxiety small and manageable. Women were encouraged to think, *If I feel anxious when I do this hard task, it's because of some silly stereotype that has nothing to do with me or my abilities.* By pointing to a cause outside themselves, women protected their identities as well as their working-memory space. They didn't expend energy questioning their abilities and so could focus on the task at hand.[74] And if you have a hard decision to make, you want every drop of your mental energy available.

So you have your first protective strategy: knowing when and how stereotype threat occurs. If you're a woman who answered True to several questions above, the next time you find yourself becoming anxious about a decision that's taking longer to make than you expected, tell yourself, *This task would be challenging for anyone. I'm anxious because of some silly societal notion that has nothing to do with me.*

The second strategy I'm about to suggest doesn't seem like it belongs here. It doesn't quite fit corporate America, most hospitals, or any courtroom. It might make you scratch your head at first, but it's been written up in hundreds of scientific journal articles as an effective way to reduce stereotype threat. Most scholars who take it for a rigorous test drive find it works.

It's a self-affirmation strategy. When most people hear the term *self-affirmation,* they think of encouraging notes that mothers put in their children's lunchboxes, or they imagine someone saying *I am lovable* or *I can do this* over and over. Research shows those kinds of positive self-statements help people who already feel good about themselves, but they can backfire for people who have low self-esteem. One study found that if you already feel bad about yourself, repeating something positive makes you feel worse, not better.[75] You can imagine the terrible inner dialogue: *I'm supposed to think I'm lovable, but I keep thinking that I'm not lovable, so I can't even do this right.*

This is not your mother's self-affirmation strategy. It's one that's based on research by multiple scientists, including psychologists Claude Steele, Geoffrey Cohen, and Gregory Walton at Stanford University and Toni Schmader at the University of British Columbia.[76] The steps are simple. When you know you have a big decision to make, first, find some time when you can think, perhaps in the morning before everyone is up or over a quiet lunch. Second, get out a blank sheet of paper and set a timer for fifteen minutes or twenty, if you have it. At the top of the sheet, write down one of your core values. The list of things you could possibly value is as large as your imagination, but to get you started, you might place a high priority on supporting your family and friends, having a healthy lifestyle, achieving financial security, leading a religious or spiritual life, working hard, learning for the sake of learning, or making the world a better place.[77] You could value something as concrete as being on time or something as abstract as being at peace. Don't pressure yourself to figure out the one thing you value above everything else; that's not where the benefit lies in this exercise. Just name something that makes you think, *Yes, that's a high priority for me.*

Once you've written down one of your values, spend the rest of your time and that page answering one simple question: Why is this important to you? Write about why it matters and who taught you that it should. Describe times when having this core value has made a difference in your life. Maybe one of your core values is supporting family and friends. When did supporting a sibling make you feel really alive? Did you cancel another commitment to be with a friend in need and find you were glad you did? You might describe a single powerful incident or a series of small ones. Just write whatever comes to mind related to your core value until your timer tells you to stop.

It's that simple, and it's surprisingly effective. What does this tactic accomplish? Writing about something that you care about helps you cope with a threat to your identity. For instance, if you're a woman who answered True to some of those earlier statements, then there's a good chance you're feeling a threat to your sense of self at work — you believe you're at risk of being labeled one of those less committed, less capable women. Self-affirmation prompts you to consider "the diverse, positive aspect of yourself," and scientists believe that this kind of reminder helps you see other events and cues in your life as less stressful.[78] You're also bringing to mind another group (besides the stigmatized group) to which you belong, a community you actively wanted to join. If you value financial security, then perhaps you count yourself among homeowners, and if you value making the world a better place, then maybe you count yourself among blood donors or soup-kitchen volunteers.[79]

Do you have to write? Would it be enough just to daydream about one of your values on your incoming commute? That might be enough for you, but researchers have found that when you write for fifteen minutes, you actively generate a vision of yourself, one that can fortify you. As one team of researchers put it, "Saying is believing," and writing helps you say it to yourself and internalize the message.[80] Besides, if you're at all like me and you try just to think about one of your values, your mind could wander and pretty soon you're rehashing a conversation from yesterday or planning how you'll spend your evening.

Here's what I find most surprising about the self-affirmation research: Because you're tapping into diverse aspects of your identity, the value you write about doesn't have to be related to the decision you face for the writing to help you.[81] If it's a work decision, you don't have to write about your career or why you chose it in the first place. You can write about why you value your healthy lifestyle instead. This isn't a sneaky decision strategy that will help you uncover the right option. This isn't even a motivational pep talk where you're trying to convince yourself to follow a particular course of action. This is affirmation. This is about reducing stress and anxiety and reminding yourself that you're much more complex than the unwelcome stereotype that threatens to overwhelm you.

Researchers were skeptical when this strategy was first described back in 1988, but in the years since, hundreds of studies have shown that this simple writing exercise reduces the fear that you're about to live up to a label.[82] By keeping that anxiety at bay, you can think more clearly, make better choices, and solve harder problems.[83] One study even showed that after completing a self-affirmation writing exercise, people were willing to reevaluate poor decisions they'd made in the past, changing them for the better.[84] All too often, we dig in our heels and defend our bad decisions, but this simple writing exercise can reverse that trend. Try it before you walk into a meeting where you'll be attempting to figure out what went wrong with a project. This activity doesn't turn you into a scapegoat, but it does help you see that you're much more than that one decision.

So can we agree to stop using the term *woman's prerogative*? It might seem funny at first, but it's outdated and it's incorrect. When people suggest that men make a beeline for their decisions but women make circles around theirs, they don't do women any favors. Once those stereotypes come to mind, it's actually harder, not easier, for women to choose well.

This tired phrase isn't the only reason that society sees women as indecisive. It's clear from the research in this chapter that if we want to understand how women decide, we need to recognize that women are often contextual decision-makers. Women are responsive to the people and cues around them, and that's a good thing. We value a man of action, but we should also value a woman of discernment.

Does this suggest that women are insatiable people-pleasers whereas men are impervious to peer pressure? Diana, the police chief, says that's the wrong conclusion. Sure, she sees young women striving to please people early in their careers, whether it's because they want to be liked or because they're trying to show they're smart enough to do the job. She admits that's how she behaved at age twenty-five. But as men and women mature, she finds that women have an easier time making unpopular choices. In her experience, women become better at being decisive "because we don't fear making somebody unhappy. Someone is always going to get bent out of shape. Not everybody gets a trophy. And I think men have a harder time making those tough decisions because they still want to be seen as one of the guys."

Men are worried about how they're seen by other men? And this is where we will turn next. We're going to look at the pressure of masculinity and see what this often-ignored issue means for men's choices. If that tension caused decision-making dilemmas for men in the workplace and men alone, it would be one thing, but we'll see how the pressure for a man to be one of the guys makes choices harder for the women in the office as well.

Chapter 2 at a Glance: The Decisiveness Dilemma

THINGS TO REMEMBER

1. There's a core belief that women take care, but men take charge.
   - Diana's story about how male and female officers approach a tense situation
   - We expect men to decide like secret agents and women like mother ducks.
2. Voters and employees alike prize decisiveness in their leaders.
3. Although many people believe that men are more decisive than women, scientists find that men struggle with their options just as often as women.
4. Women walk a tightrope of collaboration and decisiveness. Society expects women to ask for input and share the credit but simultaneously criticizes them for being too dependent on others.
5. Companies with more women on their advisory boards seek higher-quality advice.
6. Stereotype threat is the splinter of anxiety you feel when you're afraid you're about to live up to someone else's negative expectations of your group.
7. Even if you don't believe in the stereotype, you can become preoccupied with the idea that other people are judging your group based on your performance.
   - Examples: African American students taking achievement tests, women doing math, and women driving with men who believe women aren't good drivers
8. Stereotype threat is one reason that women are seen as less decisive. When women feel anxious about someone else's low expectations of them, they underperform on hard decisions.

9. Reduced working memory limits how much anyone can consider in a given moment.

10. We may savor the idea of a man of action, but we should also value a woman of discernment.

THINGS TO DO

1. It's hard to distinguish anxiety from other strong emotions. Look for cues in your environment to gauge whether stereotype threat is a concern.

2. Remember to tell yourself, *This decision would be hard for anyone. If I feel anxious, it's because of some silly stereotype that has nothing to do with me.*

3. When you face a challenging decision, use an affirmation exercise to prevent these anxieties from literally getting the best of you.

   - Write about one of your core values for fifteen to twenty minutes.
   - You'll remind yourself of who you are and keep the stress of stereotypes from interfering with your decision-making process.

3

Hello, Risk-Taker

WHEN VIVIENNE was in her early twenties, she wanted to make a movie. She and a friend started a production company and picked a clever but little-known short story as the basis for their first script. Then came the problem of funding. She and her business partner didn't have any money, so Vivienne set up meetings with people who did. Looking back, she calls it "selling sizzle." At some point over lunch with a prospective funder, she would lean in and say, "If we make the movie, we'll put your name in the credits." A small promise, but she said it often worked.

She was asking them to take a big risk. No one had heard of the story or the author. Vivienne had never produced a film. She didn't know the movie industry and she hadn't even finished college. She was enrolled at a state school, dabbling in classes. At the time, the most impressive credential on her résumé was "summer job in a doctor's office."

But her lack of experience and focus didn't seem to be much of a problem. People were remarkably willing to reach for their checkbooks and back the enterprise. After several months, she had all the money she needed to make the film.

Twenty years later, Vivienne Ming found herself seeking funding for other projects, this time for innovative educational technologies, but by then she knew what she was doing. Vivienne had become an expert, although not in film (the film plans fell through). She had gotten serious about school—very serious, earning a PhD in cognitive psychology and theoretical neuroscience. A computational genius, she landed appointments on the faculty of both Stanford and Berkeley. The Gates Foundation contacted her, as did the White House Office of Science and Technology, both seeking her advice on tech and education issues. By 2015, she had founded five different companies, each more impressive than the last, and *Inc. Magazine* had named her one of its Ten Women to Watch in Tech.

But lately, when Vivienne sits down to a business lunch, she meets with something unfamiliar: resistance. Investors are slower to take out their checkbooks. Vivienne can show a demo of her product and she can prove her previous success by quantifying how much money her other tech projects made—something she certainly couldn't do with that movie. Venture capitalists listen and ask her about potential markets and risks for each project, and—when they're going to say no—they often reassure her: "You've really done something impressive here." Vivienne says, "They 'avuncular-ize' me. Perhaps I just made that word up, but if it isn't a word, it should be. They treat me like an uncle would treat his seven-year-old niece when she's showing off her stamp collection." Vivienne still finds funders, but she has to do more proving and convincing than she once did, even though you'd expect investors would be scrambling to hitch their wagons to her star. She had once been surprised how many people trusted her with their money. Now she's surprised how many hesitate.

What has changed?

It could be the economy. After the economic crash of 2008, investors may be scrutinizing the numbers more than they did in 1993. It could be the size of the investment. Whereas Vivienne needed to raise only $100,000 to produce her small, independent film two decades ago, she needs to raise considerably more to fund a software project today. It could also be that venture capitalists would have been more interested in her at twenty-two because they tend to look for young upstarts. John Doerr, the esteemed Silicon Valley venture capitalist, once said, "I'm still looking for a bunch of burrito-eating Stanford kids who walk in with no idea what a business plan is," although even this admits a need for some credentials—he wants team members who have proven their mettle at Stanford.[1]

But I propose another explanation. I believe Vivienne meets more resistance now because now, she's a woman. When Vivienne approached funders for the movie years ago, she introduced herself as Evan. Vivienne is transgender.[2] As her father's oldest son, she lived as a male for most of her early life, despite profound depression in those roles. In her thirties, having long identified as a woman and having always wished she could live as a woman, she made that challenging transition. So when she meets with funders today, they see a brilliant, feminine woman with shoulder-length blond hair sitting on the other side of the table.

Do investors hesitate because she's transgender? "I don't exactly lead with that part of my history," she explains. "It's out there on the web so they could find it, sure. You could look up my story in the *New York Times,* the *Huffington Post.* It's even on my website if you dig into my personal page. But that's not what most investors research on a prospective project before they sit down with you." Are venture capitalists thinking, *Something is different about her,* and does that give them pause? Possibly, and no one can know everything that's running through potential investors' minds when they push away from the table and say, "Thanks, but this isn't the project for me."[3]

But as we'll see, this isn't a problem unique to transgender women.

It's a problem for all women. What does gender have to do with it? A smart investment is a smart investment, right? Regardless of who presents it. But people hesitate to take risks on women. Society sees risk-taking as a man's world, and that leads to a wide assortment of problems for women. Women often have few compatriots in risky leadership roles, so when a woman sticks her neck out, she stands out. This makes risk-taking riskier for women generally and leads many people to second-guess those women, especially when they take on roles that are usually held by men, such as founders and executives of technology companies.

This chapter explores the connections between gender and risk, how people look at someone and see a blinking light of caution or a green light of possibility. What comes to mind when we think of males and risk? How do we respond differently when we think about females and risk? We'll explore whether it's right to assume that men take more chances than women, and we'll uncover some surprising circumstances in which men feel urged to take risks when none are needed. At moments when men do put more on the line and women hold back, why are they behaving so differently?

Follow the Money

Many frustrated female entrepreneurs will tell you that they've had similar experiences to Vivienne's when they sit down with a venture capitalist; data confirms that investors are less willing to risk their money in a project led by a woman. Perhaps the most persuasive findings comes from Alison Wood Brooks, a Harvard Business School professor who studies how people respond to pitches for new startups. The pitch is an essential step for entrepreneurs who want investment money. The hopeful entrepreneur talks with a venture capitalist, just as Vivienne did, sometimes at lunch but often over a Skype call, and the entrepreneur puts her best ideas forward, trying to convince the investor that she has an inventive solution to a common or emerging problem. Brooks and her colleagues found that investors were 60 percent more likely to invest their money in a project pitched by a man than one pitched by a woman, even when the man and the woman pitched the same idea.[4] Some people might skim over that last sentence and figure, well, 60 percent is close to 50 percent, so it's nearly equal. But watch your reasoning. Equality would mean that investors were 0 percent more likely to invest in a project pitched by a man.[5]

Did the male entrepreneurs pitch their ideas more confidently than the women? We'll explore the gender issues around confidence in the next chapter, but certainly, yes, if men showed more confidence, it might have swayed investors. So Brooks ran another experiment and had professional actors give identical pitches so entrepreneurs of both genders projected the same levels of confidence. Investors once again chose to fund the men — at a ratio of more than two to one.[6]

In the real world, this means men see more money than women when they take risks. Much more. Between 2011 and 2013, venture capitalists invested $50.8 billion in new business projects, and $49.3 billion of that money went to companies led by male CEOs.[7] Of course, the majority of CEOs are male, so it's unsurprising that they'd attract the lion's share of the investment dollars. But does that mean investors would be more receptive to a risky project proposed by Evan than one proposed by an equally qualified Vivienne? Unfortunately, that's what the data suggests. Brooks's research indicates that people give men more money, and more often, to take risks.

The Sure Thing and the Risky Long Shot

Before we go any further, we should define *risk-taking*. *Risk-taking* means passing up a sure thing to shoot for an opportunity that might be costly but that also might turn out to be more valuable than the sure thing you passed up. When you take a risk, you decide you won't go for the bird in your hand. You're going for the two in the bush.

Taking a risk means buying a car online, sight unseen, to get an incredible deal. You could test-drive a comparable car at a nearby dealership, but you'll pay 25 percent for that peace of mind. Or let's say you go down to the cafeteria at lunch and some friends call out your name, pointing to a space at their table. Do you sit with them, the sure thing, or do you join the attractive new person who is eating at a table alone, the risky opportunity?

We think that some people prefer to live on the edge, but contrary to popular belief, risk-taking is not a personality trait.[8] It's not the same as being an introvert or an extrovert, a shape you bounce back to whenever you can. If you're strongly extroverted, you'll find it energizing to interact with people in most contexts, whether you're walking into a dinner party full of strangers or meeting with your favorite client. But risk-taking doesn't work that way. Research shows that people who love hang-gliding and bungee-jumping are intense risk-takers in terms of how they want to spend their free time, but when asked how they want to save for retirement, they look quite normal in their choices.[9] A person who is brazen in one context might hesitate in another. You may not bat an eyelash when you claim a few questionable deductions on your taxes, deductions that would be hard to justify if you were audited, but you might be much less of a risk-taker when your niece asks if you want to hold her pet snake.

Facing the Fireman's Pole

Americans have colorful (sometimes crass) idioms that emphasize that men, whether they like it or not, need to be risk-takers in this culture. Sayings like "Be man enough," "A man's got to do what a man's got to do," "You need to man up," "Are you a man or a mouse?," and "Grow a pair" are just some of the ways our language sticks men with all of the risk-taking responsibilities. It's interesting to look closely at these expressions. They don't suggest a man needs to take risks for the good of his family or for the love of his country; in these, a man's very manliness depends on his taking risks.

Does a woman's womanliness hinge on the same level of courage? Not exactly. Descriptions like "shrinking violet" and "nervous Nellie" suggest that, rather than expecting women to stand tall, society expects them to stand as far away from danger as possible. In fact, comparing a man to a woman is a quick way to say he isn't taking the risks he should, that he's afraid. Picture a thirteen-year-old boy who dares to say, "Wait a minute, guys, I'm not sure this is safe," to his friends. He's reconsidering a decision and wants time to reevaluate, and we can just hear the names he'll be called: a mama's boy, a sissy, a wuss (a blend of the words *wimp* and *pussy*), and any number of other putdowns that suggest his manhood is in question.[10] Our language makes it clear that it's almost the definition of masculine to embrace risk, and it's feminine to fear it.

One of the most revealing places to study how Americans think about gender and risk-taking is on a playground. Barbara Morrongiello, a psychologist who wants to reduce childhood injuries, has been going to playgrounds to watch how parents respond when their children climb to the top of a tall slide or get on a jungle gym. If women are less comfortable with peril, if women find it hard to tolerate risky choices, then we might expect mothers to be more protective of their children than fathers are. But they aren't. Across multiple studies, Morrongiello finds that the parents' genders don't affect how protective they are of the child who is playing in ways that might be dangerous.[11]

The single factor that affects how protective parents are at playgrounds? The child's gender. Parents—mothers and fathers—are more protective of daughters than sons.

In one of her studies, Morrongiello watched how parents responded when their children approached a fireman's pole for the first time and had to learn how to slide down it. A fireman's pole isn't extremely dangerous, but it's not as safe as, say, a sandbox. A fireman's pole is tall and involves taking the risk of letting go when you really want to cling tightly. Parents pressured their sons to keep trying to slide down the pole, even if the boys said they were scared, but if daughters said they didn't want to attempt the pole, mothers and fathers usually said that was fine. Even children who showed no difference in physical ability were treated differently. When girls did want to slide down the pole, mothers and fathers jumped to assist them, often keeping their hands on their daughters' waists or backs, even if the girls didn't ask for help. But parents were much more likely to just stand by and coach their sons, saying things like "Reach over and grab the pole" or "Loosen your hands a little." Little boys were encouraged more often than little girls to try it again if the first attempt went poorly.

The difference is subtle. These parents weren't calling frightened little boys mean names, and they weren't scolding little girls when they climbed to the top. Yet the gendered message is clear: risk-taking is fine for little boys, encouraged even, and with a little instruction, it won't be as risky the next time. But it's hard to let little girls be anything but safe.

Most of us don't have to slide down a fireman's pole in the workplace or at home, thankfully, but we do have to take other kinds of risks and we have to be given the space to take them. Career opportunities are affected by the quick, automatic, and often unconscious judgments that colleagues make about whether or not you can be trusted to take calculated risks. Let's imagine that you're applying for a job, one that requires other people to see you as a highly effective risk-taker. The job might involve creating an innovative product line, saving a misman- aged project, or negotiating new contracts. If you're in civil service, the work might involve making decisions that could save lives or imprison criminals.

You're asked to come in for an interview, so you do your home- work. You research the organization and identify where it's strong and where it has room to grow. If you're a woman competing against men for the job, as soon as you walk in the door, there's the problem of your size. When people are asked to judge whether or not the stranger standing in front of them will be an intrepid risk-taker, they size that person up. Literally. Taller and stronger people are rated as bigger risk- takers than shorter and weaker types.[12] It's not clear why that connec- tion is made — perhaps someone who is physically impressive looks healthy enough to take risks — but most women are at an obvious dis- advantage compared to men on height and strength. According to re- searchers, when another person is sizing up your appetite for risk, it helps if you look like you could bench-press a hundred pounds.

Then there's the issue of perspective, of who is sizing you up as a future risk-taker. If it's a male manager and you're a woman, there's a good chance you'll have more proving to do than you would with a female manager. Research shows that on a given gamble, men under- estimate by 20 percent the likelihood that women will take risks.[13] "But they have my résumé," you're protesting. "They can see that I haven't taken the easy road, that I've stuck my neck out on some risky proj- ects." Unfortunately, the facts might not be enough. At least one study shows that males' biased perception persists, that men continue to be- lieve a given woman will be cautious even as they look at data docu- menting her actual risk-taking history.[14]

Americans don't just *expect* that men will take risks while women will be cautious; it's what people think they ideally *should* do. Two

Princeton University researchers, Deborah Prentice and Erica Carranza, asked students to indicate how desirable it was for men and women in American society to possess certain characteristics. Prentice and Carranza looked at a wide variety of traits — from being dependable to being cynical, from being athletic to being shy. As we might expect, participants considered it highly desirable for women to take care of the people around them. It was important for women to be warm, kind, and interested in children, but one could be an outstanding man in American society without shining in those areas. The different emphasis placed on risk-taking for men and for women might surprise you. Out of forty-three desirable traits, being willing to take risks ranked fourteenth for men (just below being competitive) but thirty-seventh for women. To get a sense of how low risk-taking ranked for women, it helps to compare it to being assertive. Most people know, whether they've heard it from a well-meaning family member, read it in an article, or surmised it from a performance review, that assertiveness is often seen as an ugly trait in a woman, as something to be toned down. Where does being assertive rank for women in Prentice and Carranza's research? It was thirty-ninth, just slightly less attractive than being a risk-taker.[15]

We can begin to see why investors hesitate to back Vivienne but would have enthusiastically supported her when she was Evan. It's easier to get excited about Evan's risk-taking — it's what people want strong young men to do. That feels right. But a woman who is taking a big risk and asking others to take that risk with her — that's a much harder sell. Investors might find themselves feeling critical of a woman proposing a big risk, but rather than realizing that uneasiness is bubbling up because risk-taking is undesirable in women, they assume there's something wrong with her proposal. *Something doesn't feel right about this,* a venture capitalist thinks, and, going with his gut, he interprets that as a sign that he should look for another place to invest his money.

Of course, it doesn't help that Vivienne is an entrepreneur in technology, a field where male superstars dominate. When journalists list the risk-takers who have changed the face of technology, they almost always name men: Jeff Bezos of Amazon, Steve Jobs of Apple, Larry Page and Sergey Brin of Google, Mark Zuckerberg of Facebook, and Elon Musk of PayPal, SpaceX, and Tesla.[16] There are, thankfully, women who make headlines for the risks they've taken in technology, for their disruptive breakthroughs, but theirs usually aren't household names like those of the men I just mentioned. Consider Elizabeth Holmes, who is the youngest self-made female billionaire in the United States and one of *Forbes'* Top Forty Under Forty. She dropped out of Stanford her sophomore year to start Theranos, a high-tech blood-testing company that has developed a way to perform seventy tests on a single drop of blood. With no vials, no tourniquets, and using a fraction of the time and money that most labs require, Theranos has revolutionized blood testing. The next time you walk into a Walgreens for a cholesterol screening and find yourself amazed at how simple and painless the blood test is, you have Elizabeth Holmes to thank.[17] But if investors are so reluctant to take a risk on a woman, where did Holmes get the money to start her company? Clearly, launching a blood-testing lab takes capital. Did she have to pitch her idea initially to a stranger over a lunch meeting? Actually, no. Her parents gave her the money they'd saved for her college tuition.[18]

The His and Hers of Risky Leadership

It's exciting to hear about Holmes's success, but how would we respond if she had used all of her college savings on a company that failed? Would we be more disapproving of Elizabeth than we would be of, say, a man who'd dropped out of Stanford and launched a tech company that fizzled?

To answer that question, let's return to Victoria Brescoll's lab at Yale University. Brescoll and her colleagues wanted to understand how people react to long shots that fail. She asked adults to read a brief fictional news story about a leader who made a risky decision. All of the participants read about a police chief in a major metropolitan area who knew several weeks in advance that a big protest was being planned downtown. A few hours into the event, the protest got out of hand and the chief of police told the officers to take action. In one version of the story, the chief did not send a sufficient number of officers to the scene, and twenty-five people were seriously injured. The other half of the participants read a story with a happy ending. In this version, when the protesters became unruly, the chief sent in a large number of officers, and the protest continued smoothly without any further incident or serious danger.[19]

In both cases, the chief of police took a risk and did not send in officers until after the protest grew problematic. In one version of the story, the chief then made the correct decision and everything turned out fine, and in the other version, the chief made a costly mistake. Brescoll's team was interested in how the participants' opinions of the police chief's decision would be affected by the gender of the chief. Did it matter whether it was a man or a woman who made the bad call? It did. When a female police chief found herself with twenty-five injured civilians, participants judged her as incompetent. They didn't want to take away her badge, but they thought she should be demoted, that she didn't have the kind of good judgment it takes to be in charge. When a male police chief made the same error in judgment, most people weren't as critical. After reading the news story, participants answered a variety of questions about how much power, respect, and independence the leader deserved, and these ratings were combined into a single status score for that leader. When the male police chief misjudged the situation and sent in too few officers, his status score in the public eye dropped by approximately 10 percent. But when the female police chief made the same mistake, her ratings dropped by almost 30 percent. A lapse in judgment cost her almost three times as much as it cost him. When the risky decision paid off, when enough officers arrived and peace was restored, both leaders were held in equally high regard, which means a female police chief wasn't automatically seen as a poor fit; that happened only when she took a risk that failed.

Brescoll and her team looked at a variety of leadership positions. They ran a scenario in which the CEO of an engineering firm faced a risky decision and another where a chief judge of a state supreme court was in the hot seat. The results were the same: a risky decision that went awry was highly costly for the women in these roles — more costly than for their male counterparts. All three of these scenarios have something else in common: These female decision-makers were in roles traditionally associated with men. A police chief of a large city, a CEO of an engineering firm, and a chief justice of a state supreme court are all leadership positions typically filled by men.[20]

Was there a situation in which a male leader was judged more harshly than a female leader for making a bad call? There was — a male president of a women's college. In this leadership role, one most people expect to be held by a woman, men paid a price when they showed poor judgment. Were there any other highly respected professional leadership roles people associated with women? I asked Brescoll that question, and she said they'd looked for other leadership positions that were both dominated by women and seen as high status. They hadn't found any. A president of a women's college was the only one. "It's kind of depressing, to be honest," she said.[21]

Brescoll and her colleagues concluded that people find it easier to accept a poor risky decision when it's made by a leader in a gender-appropriate role. His areas of expertise are not interchangeable with hers, and leaders are more severely judged when they make a mistake in the other gender's territory. So would we be more likely to condemn a woman who poured money into a tech company that bottomed out? Probably.

This raises two serious concerns. First, men are associated with most top leadership roles: CEOs and VPs, politicians, movie directors, military officers, airline pilots, surgeons, lead engineers, and law partners, and the list goes on. So if men are typically associated with these roles, then the many women who take on these jobs are stepping into a kind of rigged game. Second, the nature of a leadership position involves facing risky decisions. Some would say this is a leader's job. Putting these two pieces together, we're left with a picture in which risk-taking is riskier for female leaders than for most male leaders. We already had plenty of reasons to believe there's still a glass ceiling — the old boys' network, which means men continue to choose men as successors; sex discrimination, which means female executives are undermined and not taken seriously; occupational segregation, which means women often hold leadership roles that don't go all the way to the top (such as jobs in personnel or public relations).[22] But now we might have another reason for the glass ceiling, one that many people don't know about, or at least don't talk about: in male-dominated fields, women who are working their way to the top fall farther than men when they take a risk that doesn't pan out.

What lesson can we learn from this line of research? If you are a woman taking risks in a role that your colleagues see as a man's job, or you're a man taking risks in a role that others believe befits a woman, you might want to do some extra coalition-building before you act on big, risky decisions. It's a prudent way to protect yourself. Find one or two powerful allies for the option you're advocating. This becomes trickier the higher up you go, as the circle of potential confidants grows smaller and smaller, but finding supporters is a wise investment of your time.

Practice Being Sixty-Forty

Is one of the lessons here that everyone should stop taking risks? No. Risk-taking is part of a fulfilling, regret-free life. Studies show that people often have deep, long-lasting regrets for the stones unturned, for the paths imagined but not taken.[23] A team of researchers at Cornell University asked men and women in their seventies, "What would you do differently if you could live your life over?"[24] Some respondents lamented their actions — for instance, "I shouldn't have married so young." But four times as many regretted their inactions, the risks they'd wanted to take but never did, saying, for example, "I should have aimed higher in my career" or "I was too meek — I should have been more assertive."

Earlier we learned that risk-taking is not a personality trait but a skill, one that can be cultivated and improved with practice. If risk-taking is a skill, how do you practice it and what should you practice? Eugenia, the chief marketing officer for a data visualization company, offers two pieces of advice. First, she actively coaches her leadership team to practice identifying where they can take risks and where they can't. She tells them, "We're growing so fast that you're juggling a lot of balls at once. Not only do you have to juggle all of those balls, but you also have to make a decision about which balls you're going to drop and how far you're going to let them roll away before you reach down to pick them up. And you're doing all of that while you're still juggling."

You can't return that phone call today? Can you let it go two days, or should it be the first thing on tomorrow's schedule? Dropping balls is inevitable, she says, and a successful leader who is in demand has to accept that risk. But if you actively practice anticipating the impact of a dropped ball and responding accordingly, you become a stronger decision-maker and a tremendous asset to the team.

Eugenia's advice about juggling is one she shares with all the members of her team, but the next piece of advice is one she reserves for women. She calls it her eighty-twenty rule. She has observed that when women are making a contribution to a project at work or giving a presentation, "The women believe they have to be eighty percent prepared and can twenty percent wing it. And by wing it, I mean trusting their judgment in the moment, going with what they already know from experience. I did a survey once of some of my friends in marketing and do you know what the men said? Men winged it much more. On average, men were comfortable winging it sixty-five to seventy percent of the time and only needed to prepare thirty to thirty-five percent of the information." So men planned only about one-third of what they would say, while women felt pressured to plan four-fifths of the points they wanted to make. Men were taking a much bigger risk getting up there because they weren't as prepared, but they were also going to speak up a lot more and spontaneously lead a greater number of meetings than women if they were willing to contribute ideas they hadn't fully researched or worked out in advance.

So what does Eugenia advise women to do? "I coach women [that] you've got to stop being eighty-twenty. You've got to start being twenty-eighty. That freaks them out, and then I say, I'm joking, but aim for sixty-forty. Don't be eighty-twenty." And this advice helps the women she coaches take more risks, giving them something relatively concrete to practice. "You'll get a lot more done and a lot more respect if you learn to be sixty-forty and learn how to wing it." What does she say to women who can't imagine going into a meeting sixty-forty? "I tell them to ask themselves two questions. Point one: Do you know more than everyone else in the room? Yes. Point two: Do you know everything that needs to be known? No. But let's go back to point one."

Don't Let Your Bosses Think You Don't Take Risks

A theme that keeps coming up in this book is that the way people see women as decision-makers doesn't always line up with women's actual decision-making prowess. And risk-taking is no different. Risk-taking is often the crucial skill that separates a manager from a leader. You can work to manage the status quo, explains Herminia Ibarra and Otilia Obodaru, two researchers writing for the *Harvard Business Review,* or you can depart from routine and be a "force for change that compels a group to innovate."[25] Male executives often think that departing from routine is one area where female executives are lacking. In their article titled "Women and the Vision Thing," Ibarra and Obodaru reported on their study of performance evaluations for almost three thousand executives from a hundred and forty-nine countries and found that female leaders received higher ratings than male leaders on almost all — seven out of eight — leadership dimensions. Women scored higher on giving feedback, on showing tenacity, and on aligning priorities, to name a few. But the single area where women fell short was considered to be the most crucial aspect of leadership — establishing a vision. Men gave women especially low scores in this area. To be seen as visionary, Ibarra and Obodaru write, you need to challenge the status quo, define new strategies, and depart from rules and routine. That is, you need to take risks.

So far, it sounds simple enough: women leaders need to stop doing things the way they've always been done. But when you ask most women executives what they do well, breaking the mold is one of the first things they mention. A study by Caliper looked at the personalities of women executives at the level of vice president and higher. Of the different personality traits Caliper surveyed, these women leaders scored especially high on risk-taking. They earned their lowest scores on "following existing rules" and "being cautious."[26]

So if you ask the male executives, they say women are playing by the rules and are anchored in the status quo, and if you ask the women executives, they say they are rule-breakers, throwing caution overboard. Who has it right? The answer could lie in how people define risk-taking. Perhaps when women leaders break a few rules, they pat themselves on the back for being visionary, whereas men believe that a leader needs to reinvent an industry entirely to earn that label. It could also be what Kat, the former tech CEO, observed earlier; in her experience, women tend to point to market research when proposing an idea, whereas men tend to claim it grew from their own creative visions.

But I think there's another key component grounded in what we notice and remember. Male colleagues and superiors might not spot women making risky decisions because they have strong beliefs that men, not women, are the ones who take chances. We tend to pay attention to and remember examples that confirm our beliefs about the world and blindly glide right past the contradictions. If you can't stand tiny dogs because you believe they bark incessantly, you'll probably make a mental note of every Chihuahua that yips at you and tell stories about that agitated miniature pinscher you saw outside a café, but when a German shepherd, beagle, or bloodhound lets loose, you're more likely to forget it, and when a Pomeranian simply blinks at you as you walk by, you won't think twice. We collect confirming instances and unwittingly ignore the rest. Scientists call this a confirmation bias, and it's an unconscious, "one-sided case-building process."[27]

Of course, confirmation bias extends well beyond stereotypes of dog breeds. People who fit our stereotypes snag our attention, so if we believe that men steer toward risk, we'll be quick to recollect our male coworkers who volunteered for risky assignments and let slip from memory the men who say, "Don't look at me." Julie Nelson, an economics professor at the University of Massachusetts, Boston, an expert on risk-taking and gender, argues that the bias to see men as the risk-takers is so strong in American society that it even compromises the way economists see their data. She points to many studies where there was iffy evidence of a gender difference in risk-taking but the scientists conducting the study drew from it robust claims that men take more risks than women.[28]

So if you're a woman who wants to advance in a role where men dominate, should you draw attention to the fact that you take risks? Or should you keep quiet in case your risks meet with negative results and put you in the position of being judged not just for the bad decision but for failing in a man's spot? Julie Nelson said one piece of advice she would offer to professional women is "Don't let your bosses think you don't take risks."[29] Take credit for the risks you've taken that had successful outcomes. Researchers at Catalyst found that the career-advancement strategy that made the biggest difference for women, the single practice that led to more promotions, higher salaries, and greater career satisfaction for female professionals, was calling attention to their successes.[30] Catalyst looked at drawing attention to all kinds of accomplishments, not just risk-taking, but successful risks should certainly be included.

Most important, point out that you're willing to take risks in the future. In *What Works for Women at Work*, Joan Williams and Rachel Dempsey argue that men and women are evaluated differently on the job: men are judged by their future promise and potential, while women are judged by their past accomplishments. It's that "prove it again" pattern, and it comes down to women being told, "You're talented, but you need a bit more experience," whereas men are told,

"You're talented and we really see you going places."[31] In their interviews with 127 successful professional women, Williams and Dempsey found that 68 percent of them had experienced the "prove it again" bias at least once in their careers. Williams and Dempsey didn't interview men, so it's not clear how often men hear the same message, but there's other evidence that women have more proving to do than men. According to a 2015 stody, on average, female CEOs have more education and more job experience than male CEOs, and female CEOs are more likely to be promoted from within, suggesting that, to make it to the top level, women have to spend more time proving themselves, not just generally, but to a single employer.[32]

Women often believe that people will like them less if they boast, particularly if they boast around someone who hasn't accomplished as much.[33] Part of that belief comes from how women are socialized — they're told not to brag — but part of it also may be cold reality. Women who boast about their accomplishments aren't liked as much, by men or other women, as modest women.[34] But in one-on-one meetings with your supervisor, make him or her aware of what you've done. Self-promotion can be a bit of a Catch-22 for women, as we'll see in chapter 4 on confidence, but it's important that women start saying, "I've taken smart risks in the past, it's a skill I've learned, and you can expect me to continue doing it in the future."

If you're a woman who's new to discussing the successful risks you've taken, start by identifying the relatively visible ones, the risks that had quantifiable results. If you pointed out a crucial flaw in a proposal that saved your group time and money, draw attention to it. Perhaps you pushed for an event that other people initially dismissed but that became the most highly attended event of the year.

Many women take risks that their supervisors don't know about or might forget, and they shouldn't be shy about pointing these out too. Early in my consulting career, I noticed that one of my new clients was late to our first two meetings. About midway through our second meeting, I took the risk of asking her, privately but directly, whether tardiness was a problem she experienced outside of our appointments. She stopped talking. I was treading on delicate ground—I didn't know her or her work patterns and she could have taken offense. She was the first person from her group to consult with us, so I risked losing not only her business but also any hope of working with her colleagues. Instead of getting angry, she hung her head and said, yes, this had become a recurrent and embarrassing problem. The two of us discussed how it might be affecting perceptions of her credibility, and over the next few meetings (all of which she got to on time), we identified concrete changes she could attempt to make. Five months later, she sent me an e-mail. Thanks to our work together, she had tackled her tardiness problem. She was more successful in meetings, and she was singing the consulting group's praises, referring several new clients to the organization. It didn't occur to me at the time that I should point out to my boss that I'd taken this risk, and so the benefits to both my client and our office went unnoticed.

This isn't typical career advice, I'll admit. No one has ever asked me on a performance review or in a job interview to describe the risks I've taken or how those risks turned out. And men don't need to prove that they are risk-takers because, by being male, it's taken for granted. But every woman needs to point out her ability to take risks when she's drawing attention to her many achievements.

What Stock Investing, Health Practices, and Extreme Sports All Have in Common

We've already seen that people are more comfortable taking a chance on men. But do men actually take more chances? Employers often look for people who take risks. Businesses that are trying to grow (or simply stay alive in a changing market) want managers who "thrive on high-stakes challenges," "are eager to apply their skills in new ways," and "are comfortable making decisions that affect multiple stakeholders," to cite a few examples from online job listings. These are all different ways of saying "We want someone who takes risks."

So *do* men actually take more risks than women? Let's turn to what is probably the most commonly cited investigation in the English language of gender and risk-taking. In 1999, James Byrnes, a developmental psychologist at the University of Maryland, led a team that carefully paged through a hundred and fifty research studies on gender and risk-taking. According to the data Byrnes compiled, men took more risks than women in 60 percent of the studies.[35] In the other 40 percent, either women took more risks than men or the two sexes took an equal number of risks.[36] So six out of ten studies argued that, yes, men do take more risks, but four out of ten challenged that claim.

We have to ask, though, whether there is any kind of pattern or if this could be random. Are there times when we can expect men to be unflinching and other times when we can expect women to be more fearless? Many researchers say yes, the sexes are predictable — it all depends on the risk they're being asked to take.

If we look at social risks, women rise to the fore.[37] What's a social risk? Taking a risk in a group; for example, speaking your mind about an unpopular issue, openly disagreeing with your boss, or admitting your tastes are different from the people around you.[38] Women are also more likely to make a complete career change, which is considered a social risk because it means leaving one's place in the organizational chart to start all over again.[39] Women are also more likely than men to self-disclose information; telling someone you made a mistake may not seem as risky as jumping out of an airplane, but it opens you up to being judged and rejected, real risks for all of us.

Women in male-dominated careers also face social risks where their male colleagues don't. When I asked Diana, the police chief we met earlier, about risk-taking on the job, I thought she'd describe the heroic risks associated with police work: being assaulted during an arrest or shot during a drug bust. But Diana says most of her daily risks don't involve a gun; they involve speaking up. She works in a large city where there is a commissioner at the top and several police chiefs who oversee different bureaus, but she's the only woman at her level. When the top brass gather for their regular meetings, there's the commissioner (male), Diana, plus five other police chiefs (all male). There might be one other woman in the room, the legal counsel, but her role is very different from Diana's. "As the only female commander, I take a lot of risks in those meetings. Every time I speak up with my ideas or opinions, I risk that the men are going to look at me like I'm an alien, wondering 'Why are you opening your mouth?'" Did Diana get those looks often? "It doesn't happen a lot, mind you, but in a male-populated environment, women still feel that pressure, that possibility." Madeleine Albright, the first female U.S. secretary of state, made a similar observation. "Early in my career, I went to numerous meetings where I was the only woman present. I would want to contribute to the conversation but would think, if I say that, everybody will think it's really stupid."[40] For a man to take a risk when he speaks up in one of those meetings, he has to say something risky. For a woman to take a risk when she speaks up, she just has to say something.

And, of course, for women, there's the social risk of being a single parent: 77 percent of single parents in the United States are women, and raising children without a partner means taking on the risks of poverty.[41] In every developed country, from Sweden to Turkey, Japan to Mexico, single-parent households might as well be called "single-mother households."[42] One woman I interviewed said, "People think women take fewer risks than men? Seriously? Those people must not know many single working mothers."

Where do men take more risks? If we look at recreational risks (or what officials call "health risks"), such as drinking too much alcohol, having unprotected sex, driving over the speed limit, swimming in unsafe places, and engaging in extreme sports such as skydiving, we find more men.[43] This kind of bodily rush, the excitement that comes with being impulsive and chasing new sensations and experiences, is what many people picture when they talk about risk-taking.[44]

Then there's the question of money. Is one sex more inclined to speculate and put their money where the risks are? Two University of California economists, Gary Charness and Uri Gneezy, looked at patterns in strong economies across the globe, from Germany to the United States to China. In the fourteen studies they reviewed, they found that, on average, men put more money on the table than women when they took a gamble or made a risky investment.[45] By looking at different cultures, Charness and Gneezy joined a chorus of researchers arguing that men and women — at their core, at their essence — are fundamentally different when it comes to risk-taking.

So according to this view, the only place where women are braver than men is in social settings, and as we've seen, that's sometimes by necessity. For many scientists, that explains the patterns in the data. End of story.

Except, of course, that's not the end of the story. Or at least, if it is the ending, it's a very unsatisfying one. I contacted a number of researchers who didn't seem satisfied with it either.

I started by asking how big the difference between male and female patterns of risk-taking really is. This led me to the economist we met earlier in the chapter, Julie Nelson. Nelson has reanalyzed the raw data from dozens of prominent research studies to get a more practical understanding of how frequently men and women take risks.[46] She takes a different approach and calculates something called effect size, which is a standard tool in statistics but not one that's used in most of the studies on risk-taking. To understand Nelson's findings, imagine creating a lineup of women. At one end we have the women who engage in a lot of risky behavior. These could be women who take risks regularly, the police officers and airplane pilots of the world, as well as the women who take hefty but less frequent risks, such as a woman who goes to Las Vegas and puts two-thirds of her chips down on the table in a single bet. At the other end, we have women who engage in very little risky behavior. These could be the women who take the fewest number of risks (like, say, your aunt who never paints her house because she's afraid she'll pick the wrong color) as well as the women who take tiny hops with each risk (perhaps your friend who wants to invest in the stock market but doesn't want to put down more than twenty-five dollars). Somewhere in the middle of that lineup we have our average woman. If she's in the middle, 50 percent of women will take more risks than her and 50 percent of women will take fewer risks.

Now imagine creating a similar lineup of men. What percentage of the men in this lineup would take more risks than that average woman? Would it be 70 percent of the men? If it's the essence of men to be audacious risk-takers, we'd expect the proportion to be high. Maybe 80 or 90 percent of the men?

Not even close. Nelson finds that only 54 percent of men take more risks than the average woman. That means 46 percent of the men take fewer risks than the average woman.[47]

Some people might look at those numbers and say, "Well, it might be a tiny difference, but it's still a difference." Technically and statistically, yes. But practically speaking, it's not much of one. If you're trying to decide which candidate is the right one for a job that involves taking risks, and you have two qualified applicants, a man and a woman, what does gender tell you about which person will go out on a limb and which one will go with the tried-and-true? It tells you very little.

Reaching Your Imaginary Arm into an Imaginary Urn

So if most men don't consistently take more risks than most women, when do men and women go in opposite directions on risk-taking? What entices men to take a gamble and what lures women into going with the sure thing?

When men and women are novices at any activity, men are likely to make much riskier choices. Many conclusions about risky decisions are based on laboratory experiments, where people read a scenario they've never seen before and decide, hypothetically and on the spot, what they would do. An example of a scenario that many researchers use in the lab involves drawing imaginary balls from an imaginary urn. Yes, an urn. If you're participating in one of these experiments, you're told to imagine an urn filled with thirty balls, ten white and twenty yellow, jumbled together. You get to play only once, you can pull out one ball and only one ball, and you can't look when you do it. (As long as you're imagining the urn, go ahead and imagine yourself blindfolded.) If you pull out a white ball, you immediately win a hundred dollars, but if you pull out a yellow ball, you go home with nothing. Now that you know the rules of the game and you know how much you could win, here's the question: What is the largest amount of money you'd be willing to pay to play?

In this hypothetical decision, men are willing to pay a lot more than women to take the risk. One recent study of 160 people found that men were willing to pay, on average, about $16.43 more. So if women were willing to pay $5.00 to play this game, men were willing to pay $21.43.[48] In a new and unfamiliar situation, men are willing to take much bigger risks than women.

When researchers ask professionals about topics they know well, not about yellow balls and imaginary urns, differences in risk-taking between men and women often disappear. For example, researchers approached financial managers in the United Kingdom and asked them to evaluate a potential contract, curious whether the managers would recommend a client to senior management.[49] There were no differences by gender in the professional risks that managers were willing to take with the contract, but male managers did take more risks in their spare time, in their hobbies and what they did for fun. Another team of researchers looked at the choices of two thousand mutual-fund investors in the United States. They found that for novice investors, people who knew very little about mutual funds or how these funds differed, men took more risks with their money than women. But when they looked at investors who had a strong knowledge of how mutual funds worked, the gender difference vanished.[50]

Of all of the studies I've read, my personal favorite was created by an enterprising team of researchers who went to the North American Bridge Championship in Boston to observe how men and women played. Bridge, sometimes called contract bridge, is a card game with a goal that sounds simple enough — collect as many winning hands, or tricks, as you can. But winning the game involves a great deal of strategy and shrewd bidding. Did the men playing in this championship make bolder bids on each hand? Did the women use safer strategies to stay in the tournament? No. Men and women competitors took an equal number of daring gambles in how they played. There was no difference in their risk-taking in this area, where they were experts. But when the researchers pulled the players aside during a break, sat them down, and taught them a new game where each player could hypothetically gamble up to two hundred and fifty dollars, the men risked 70 percent more money than women.[51]

Research studies in real-world contexts aren't nearly as common as controlled experiments. It's much easier to ask random people about an imaginary urn than to show up at someone's workplace and ask an expert to evaluate a contract. And you can control for variables in the lab that you can't control for in the stock market. But in real-world studies, there is a growing pattern: If you ask people about a hobby or something they're seeing for the first time, men are likely to make the riskier choices. But if you approach professionals with experience and ask them about something to do with their jobs, this difference between men and women disappears. Isn't this what we'd expect? If you're a professional, someone who is paid to make decisions, you have to be able to differentiate between the risks that are worth taking and the ones that are exciting but potentially reckless. And that applies to men and women alike.

Even among professionals, however, there are costs for women in visible leadership roles who make risky decisions and who see those risks fail. We've noted that a woman is more likely to be penalized for taking risks when she's in a leadership role that society sees as a man's job. If she's in a career that's dominated by men, she'll probably find that people are highly critical of the first risky decision she makes that doesn't work out. The next time, consciously or not, she might not be so quick to roll the dice. It's going to be harder for her to practice her risk-taking skills than it would be for a man in that role.

Ping-Pong Tables and Pink Lotion

The claim that men and women take an equal number of risks in professional settings will not sit well with everyone. Most of the women I interviewed who worked in male-dominated fields, such as technology and law, said that they saw their male colleagues taking more risks than women. Some of them said it was the men who pushed for an unconventional new product design while the women on the team wanted more data before diving in. Some said that they could count on more of the men to ask for outrageous terms in a negotiation and never flinch. And these women were describing experts in the field, so we can't chalk up the difference in risk-taking to inexperience.

It could be that people don't remember the times that men who are geniuses turned down risky ideas because this memory doesn't fit their expectations of risk-seeking men, and all of us tend to forget encounters that don't fit our expectations.[52] But if you're someone who sees men taking more risks, there's another explanation: the cues in your work environment might encourage fearlessness in men and discourage it in women. I'm not suggesting that there's a sign anywhere in your office that reads *Men, go forth! Women, sit down.* Nothing that blatant. Researchers are finding that men and women respond to the subtle cues, and the same cues that egg him on can make her hesitate.[53]

Stanford University social psychologists Priyanka Carr and Claude Steele looked at two such cues that can sway risk-taking. They asked men and women to participate in what can best be described as a very, very modest lottery.[54] Players chose between a good chance of winning a tiny jackpot (say, an 80 percent chance of winning one dollar) or a poor chance of winning a jackpot that was three or four times larger (for instance, a 20 percent chance of winning four dollars). It's like the decision people make when they go to the minimarket to buy lottery tickets. They can buy scratch-off tickets, where they'll have a pretty good chance of winning but the prizes will be small, or they can spend that same amount on the national lottery, where their odds are extremely low but if they do win, they'll win big. Four dollars per round was hardly big money by any standard, but participants still had to decide whether they wanted to go for a bird in the hand or two (or three or four) in the bush.

Much like the imaginary-urn experiment, the lottery that Carr and Steele used is a common scenario among researchers who want to understand risk-taking, but these scientists did a clever thing that no one else had tried: They varied how they described the decisions. They asked half the participants to check a box indicating their gender and then told them that they were about to make decisions that would measure their "mathematical, logical, and rational reasoning abilities." We'll call this the math group. This sequence of instructions — please fill out some paperwork, tell us your sex and age, and this will be a test of your reasoning abilities — is the standard way many decision-making experiments begin.[55]

The other half of the participants started the experiment in a slightly different way. Carr and Steele changed a key sentence in the instructions: They told men and women the tasks they were about to complete were "puzzle-solving exercises." No mention of math whatsoever, and the experimenters didn't ask about gender until after the experiment was over. As you might recall from chapter 1, when women aren't reminded of their gender when they sit down to solve a math problem, they're less mentally preoccupied with the performance baggage that comes with being a woman who has to prove her math skills.

Did the small cues about gender and what was at stake change the risks people were willing to take? The cues did have an impact. For the math group, we saw what economists have seen in the many studies of financial risk-taking: Men made more high-risk choices with the money they were given at the start of the experiment than women did. In the math group, men chose the bigger, unlikely gambles twice as often as the women. Men didn't choose big gambles every time, but they did choose them most of the time. To illustrate the point, imagine that you ask a group of people in your office to evaluate ten risky proposals. If it's a group of men, they would approve eight of the high-risk proposals; if it's a group of women, they would approve only four. So a man who was cued to think, *I'm a man who is being tested on math,* and a woman who was cued to think, *I'm a woman who is trying to prove she's skilled at math,* fit snugly into the molds of bold men and cautious women. It's as though when people are told their math abilities are being tested, men puff up their chests and women chew on their pencils.

But participants broke out of those molds when they were cast as puzzle-solvers. Women who were focused on solving puzzles and hadn't been reminded they were women chose the larger, riskier payouts just as often as the men. Without any negative stereotypes clouding their thoughts, those women made more gutsy choices. Perhaps equally telling? Men's risk-taking also grew or shrunk based on how they thought of themselves. The men who saw themselves as "men doing math" took more risks, 25 percent more, than the men who saw themselves as "people doing puzzles." If you presented ten risky proposals to this puzzle-solving group, you'd expect the men to approve six and the women to approve six as well. Taking away the gendered framing leveled the risk-taking playing field.

Is the real world more like the math group or the puzzle-solving group? Few, if any, professionals start their workdays by asking, "Would all the men please raise their hands? Great, now all the women? Just remember, folks, we're here to do math." But before we dismiss Carr and Steele's work as irrelevant to the real world, let's consider a second study that looked at how cues affected risk-taking. This study was conducted by Jonathan Weaver, now a psychologist at Michigan State University, and his colleagues Joseph Vandello and Jennifer Bosson at the University of South Florida. Each young man who came into Weaver's lab sat down in front of a tray of random objects—such as a power drill, a tube of toothpaste, a flashlight—and was asked to assess the usability of one of the objects.[56] The experimenter handed half the men the heavy power drill and asked them to close their eyes for ten seconds, think about how the drill felt, and describe the features that enhanced or detracted from the power tool's usability. Then the experimenter set the drill aside and said that the subject could now win some money in a dice-rolling game. After giving each man five dollars of gambling money, the experimenter explained that every time the participant rolled the die, he could bet on whether it would be an odd or an even number. A one-dollar bet meant he'd either win a dollar or lose a dollar. Each person could keep whatever he earned or held on to. The men were told that they were being videotaped to make the betting feel more public.

The experimenter gave the other half of the men the exact same instructions about the dice-rolling game, but instead of testing a power drill, these men tested a bottle of fruity-smelling hand lotion in pink and lavender packaging. These men were supposed to squirt some lotion on their hands, close their eyes, and spend ten seconds thinking about what made the lotion more or less usable. Why the lotion? Previous studies had shown that slathering up with strongly scented sweet-pea lotion was an effective threat to most heterosexual men's sense of masculinity. About half the men said it made them feel demasculinized. Weaver told me that it's rare to get men to admit they feel anything but manly, which suggests the lotion worked.[57] (You might be wondering about gay men, but they weren't included in this particular research study.)

So half of the men held a heavy power drill and half slathered on a fruity-scented lotion, and then they all played a gambling game. What's remarkable is how much men's risk-taking behavior changed when they felt their masculinity had been threatened. Most of the men who'd tested the fruity-smelling lotion bet more money in their first two rolls of the die, 30 percent more, than the men who'd tested the power drill. The lotion-testing men also bet the maximum amount more often. In essence, the masculinity-threatened men looked for the riskiest thing they could do right away, and they did it not once, but multiple times.

Some women have read this and said, "But I know my husband

would be fine with the lotion. That wouldn't change anything for him." Will all men respond this way? Of course not. The average for a group doesn't dictate the behavior of any single individual. There will be men who show more of this effect and men who show less, but the overall pattern illuminates an important gender difference.[58] Weaver has been studying the notion of manhood for almost a decade and his conclusion is "manhood is precarious — it's elusive and tenuous."[59] Men don't establish it once and then move on. They feel a need to establish their status as men over and over. When the video camera was rolling, Weaver explained, and the men were slathering on the lotion, they worried that other people wouldn't see them as real men. Weaver and his colleagues hypothesize that when straight men feel that their manhood is on the line, they're inclined to look for a way to prove themselves, to ensure their manliness isn't in doubt. And in American culture, showing that you're willing to take a risk is considered manly. Their behavior seems to echo the same message we hear in the taunts "Man up," "Don't be a sissy," and, for a direct reference to testicular fortitude, "Nut up." If your masculinity is in question, take a risk and all will be restored.

Is womanhood just as precarious? No. Most people assume that womanhood is a biological milestone, something that a girl passes into as she matures. In fact, what would it mean for a woman to no longer be a woman? Weaver, along with his colleagues Vandello and Bosson, found that most people thought that the sentence "She's no longer a woman" meant that a woman had had gender-reassignment surgery, whereas the sentence "He's no longer a man" didn't imply a biological change.[60] It could mean surgery, but most people, both men and women alike, thought it meant a guy wasn't living up to society's definition of what a real man did. So a man has a strong incentive to take risks, one that a woman doesn't share.

Most of us, male or female, don't have to worry about someone forcing fruity hand lotion on us before we walk into a meeting. But it's intriguing to put the lotion study next to the math study to see what they suggest. The implication is that men feel a need to prove their manhood in a variety of circumstances. When men are tackling a challenge where men are expected to excel—such as math—they take more risks. When they've had their manhood threatened and they need to earn it back, again, risk-taking is the answer. For men, taking risks is a way to prove their standing as men. Not smarter risks, necessarily, just bigger risks and more of them.

This notion that the environment goads men's bravado and that taking risks is a popular way to prove one is a real man is especially concerning for women who want to work at startups. The culture at startups can be, as we heard from Jessica in the introduction, like going to work every day in a frat house. I spoke with Joelle Emerson, a gender-discrimination attorney and the CEO of Paradigm, about physical spaces. Paradigm is a strategy firm that advises businesses, including technology startups, on how to create diverse and more inclusive organizations. As a discrimination attorney, Emerson used to help one woman at a time; now she's trying to help organizations treat all women well, all of the time. Company executives often ask her to come by the office to discuss why they can't attract more women to work for them. As Emerson walks through, she notices the foosball table, the Ping-Pong table, and the keg. "People don't think about the kinds of messages they are sending when they set up their physical space," she said.[61] Kieran Snyder, the CEO of Textio, heard similar observations when she interviewed women who had left the technology industry. One software engineer didn't complain about the masculine physical space but about the masculine verbal space: "Everything in our office revolved around video games and beer. We were an accounting software company, so you might think that most of our conversations were about accounting or software. But no, mostly video games and beer."[62]

You can see how these work environments might make some women feel unwelcome. But now we've uncovered a new problem with these spaces: they also spur on men to take more risks. Risk-taking is valued in startups. Most startups are trying to do something that hasn't been done, innovating a service or a product, and hoping to make a huge profit doing it. The risk-takers are celebrated, promoted; they get the big bonuses.[63] And if there are video games, foosball tables, and kegs as regular reminders of masculinity, we can expect the men will stand out as the risk-takers. It sets up a dynamic in which many women will be reminded that they're women (or at least not men) on a daily basis, which means they will be discouraged from standing out or taking risks and will also have to work extra hard to catch up with all the risk-taking the men around them feel compelled to do.

A Safer World and a Safer Dating App

There's another way to ask the question about risk: Who feels safe and who feels anxious? If someone feels safe, if someone feels protected from harm, we'd expect that individual to be more comfortable taking chances because loss isn't a real possibility. Likewise, we'd expect those who feel anxious and unsafe to take fewer chances. If you're out having drinks with two close friends and you tell them a story that you think is funny but it turns out they don't, it's no big deal. Even if one of your friends is offended, the friendship won't end. You'll work it out. But if you're meeting your future in-laws or having dinner at your boss's house, you tell only the stories you've told before, the ones you know won't make the room fall silent.

Let's start by thinking about who generally feels safe. Researchers have observed that it's specifically white males who feel the world is a safe place, and as a result they have the lowest perception of risk. Economists even have a name for it: the white-male effect. White men find many aspects of the world less risky than the rest of the population does. They see fewer risks around climate change, gun ownership, toxic-waste sites, natural disasters, motor vehicle accidents, even sunbathing.[64] One national survey of 1,489 people found that, at least in the United States, about 30 percent of white men think the world is a relatively risk-free place, and their extreme perceptions bring down the average for all white men.[65] Unfortunately, most scientists who study risky decision-making don't specifically look at race, but those who do include it in their analyses find that whether a man is white or not has a huge impact on whether the world feels like a safe place to him. Typically, African American, Latino, and Asian American men think the world looks just as risky as white women and women of color do.[66] The conclusion these researchers came to is that people are missing the point when they compare men's risky decisions to women's. They should really be comparing white men's risky decisions to everyone else's.[67] As one team of researchers put it, "Perhaps white males see less risk in the world because they create, manage, control, and benefit from so much of it."[68]

Setting tense meeting rooms and toxic-waste sites aside, Susie J. Lee sees another area where the world feels safer for men than it does for women: dating websites. Susie is a digital artist and the CEO of Siren, a dating app designed by women with women in mind. I met Susie in her favorite coffee shop, Ada's Technical Books, in a hipster neighborhood of Seattle. We took a table that was actually a thick piece of glass mounted atop hundreds of small toy compasses that jiggled as we sat down. I couldn't help but notice the symbolism—we were set to talk about men and women finding one another, and here were all these little arrows settling into slightly different directions. Some of the compasses were obviously broken, but it gave you the sense that even if we all say we have the same destination, we might follow different paths.

I asked Susie what a dating app designed "with women in mind" meant. Susie handed me her iPhone and suggested I give it a try. Siren's interface looked like an art exhibit, all black and white, with a font that was simple and inviting. To launch an account, users typed in their names and ages and uploaded a photo. Then they got to answer Siren's Question of the Day, something like "What's your favorite sandwich?" or "What nickname did you want to be called when you were a child?" Susie explained that the hope was to create a chance for users to banter. (At the time I used the app, in September 2014, all of the banter was between men and women, but it opened to the LGBTQ community in late 2015.) As I scrolled through men's answers to the sandwich question, I was surprised by how much they revealed. Some women might take an interest in the paleo guy who said, *I don't eat bread anymore, but I love a lettuce-wrapped bacon turkey burger,* and others might like the philosophical type who answered, *Discipline, play, discipline.* Women who were turned off by both of those comments might be drawn to the guy who wrote, *Give me a PBJ and I'm all smiles.*

So far, Siren's questions were much cozier than the ones I'd seen on some other sites (for instance, "What's your income?," "Do you feel unable to deal with things?," and "Which is worse, starving children or starving animals?").[69] But nothing yet screamed *This app is designed with women in mind.* Women might be amused by the sandwich question, but so would most men. (I made a mental note to ask my husband what he'd wanted to be called when he was a kid. I was pretty sure it was Superman.)

As I scrolled through the profiles on Susie's phone, I noticed that many guys had a prominent blue *V* next to their name. "That means I've made myself visible to them," Susie said, leaning in. "Every guy that I marked with a blue *V* can see my photo." And this is where the app is tailored to women: Women make the first move. When a woman taps the *V* next to a guy's name, her photo appears in the app on his phone. Women who use Siren can see every guy's photo and his answer for the Question of the Day. But men have a very different experience; they can't see a woman's photo until she says so. Initially, men can see only the women's answers to the questions. The man still gets to decide if he's interested in following up with her, but she decides whether that's even a possibility.

"A growing number of people like the idea of using a dating app but they don't like the actual experience," Susie explained. "And women are the least thrilled. You set up your dating profile where you take a lot of time to describe your interests, what you like to do, and what you're looking for right now. You post a photo. If you look really hot in that photo, then thirty guys immediately say they're interested. You're thrilled, at first, but then you sift through the responses and you're not sure most of them noticed anything but the picture. And then you get a really creepy response, the guy who won't leave you alone. It might only be one guy, but you don't want to open the app anymore."

The national data support Susie's analysis. At least in the United States, 42 percent of women who have tried using a dating app or website have been contacted by someone "in a way that made them feel harassed or uncomfortable."[70] It happens to men too, but much less often — only 17 percent of American men who use dating apps have dealt with an unwelcome advance. (The researchers didn't ask about sexual orientation, so it's hard to know if the harassment rates are different for gay and straight men.) So women have a greater risk of being harassed than men when they're using a dating app, but as Susie points out, if you're a busy professional looking to meet someone outside of the office, what option do you have?

Susie's hope is that by letting women do the screening, they will receive less harassment. And if a guy does make a woman uncomfortable, she just taps a button and makes her photos not visible to him again. So far, this structure is working: Siren has over five thousand users but zero reports of harassment, and numerous users have commented that the creepy dynamic they've had on other apps doesn't arise with Siren.[71] For women, it's one less risk they have to face when they sign up to meet someone.

So here's the real test: Are 4,990 of those 5,000 users women? I can't help but wonder if guys are going to use a service where they can't see pictures of the women. Susie had also wondered about this, but one year after Siren launched, 45 percent of its users were men. There are still more women than men, but it's not the gaping difference I expected, and it's better than some national dating websites, where 68 to 72 percent of the dating profiles are women's.[72] "A good number of men out there are fed up with standard dating sites," Susie explained. "I just read an e-mail this morning from a guy who said he's used OkCupid in the past, and even though he expressed interest in over a dozen women, he never had a single woman reply. None. Crickets out there. But he's been on Siren for two weeks, and he's already had two women say they'd like to get together." Men might also feel chosen in a way that's appealing.

What can we learn from the white-male effect and the appeal of dating apps like Siren? One lesson is that we don't all feel the same level of safety. Women have more reason to be apprehensive about dating websites than men do, and African American males have more reason to worry about gun violence than their white peers. Do feelings of safety change the risks we feel emboldened to take? Absolutely. When you feel you have a safety net, it's much easier to take a chance. Another lesson from the Siren website experiment: If we give women more control in an area that ordinarily feels risky to them, if we truly reduce the level of risk involved for women, they like the experience much more. But Siren didn't reduce risk-taking by making things equal for men and women. They reduced women's sense of risk by enforcing different rules for men and women, by giving women more control than men had. It's an interesting potential lesson for any company that is desperately trying to hire and retain more women.

But How Much Could Speaking Up Cost Me?

So who feels anxious? Susan Fisk, now a sociology professor at Kent State University, has been studying how much anxiety men and women feel when they face a risky setting at work.[73] What's a risky setting? A risky setting is a social situation at work in which you're with at least one other person and there's a safe way to go and a risky way to go, and you can't tell in advance whether the risky way will cost you or benefit you. If you're in a meeting with judgmental coworkers and you have a new idea, you could go the safe route and not mention it all. No gain, but also no harsh judgment. Another alternative would be to share your idea over lunch with one of the supportive people on the team. It's a safe way to test the waters, but you might not be credited with the idea, if the idea even makes it back to the group. Or you could go the risky route and raise the idea in the meeting, perhaps inviting ridicule (or however your team sanctions people for bad ideas) but also having the opportunity to improve your standing within the group, to be noted for your good idea, and to bring some innovation to the table. Another risky setting comes when your boss calls you into his office and asks you to give him feedback on a presentation he just gave. You could go the safe route and simply point out the strengths, or you could go the risky route and identify the problems. If you go the risky route, he could be incredibly grateful for your honesty—he gets a chance to improve the presentation before he gives it again—or he could be defensive and angry, viewing you as someone who isn't a team player after all.

Fisk asked adults to imagine themselves in several scenarios like these, rate how risky these situations were, and write about what they would do and how they expected they would feel. When the scenario was low-risk—such as speaking up in a meeting with supportive colleagues—men and women both felt very little anxiety. But make that scenario risky—put someone in the room who shoots down every idea and says things like "Does anyone actually have a good suggestion?"—and women experienced a spike in anxiety and negative feelings. Men fully agreed that speaking up in the latter scenario would be risky, but it didn't bother most of them in the same way. A handful of men felt a spike in anxiety, but most men felt the same moderate level of anxiety whether the work situation was risky or not. And it's that difference in anxiety, Fisk finds, that makes all the difference. The women and men who experienced greater levels of anxiety about the situation were less inclined to take a risk. But it was mostly women who experienced that heightened anxiety, which means the women were taking fewer risks.

What were these women so anxious about? Was their anxiety driven by a fear of the worst possible outcome, that they might be demoted or fired? Did they think they had a high chance of saying something truly stupid?

When Fisk looked closely at the data, it turned out that these women weren't afraid of failing; the factor driving their anxiety was the concern that they might not experience resounding success. Anything less than 100 percent success and the risk wasn't worth taking.

Why might this be the case? Fisk argues that because women are held to higher standards than their male peers, they have to perform at the very top of their game—better than men, in fact—to receive the same credit. For risk-taking to be "worth it," women have to feel that they have an excellent chance at a stellar outcome.[74]

What do I make of these findings? Researchers already knew that women were less likely to speak up in meetings and less likely to apply for new jobs that might be a stretch. But what they didn't know, or at least didn't fully recognize, is that it's common for women to feel that they have to achieve the best possible outcome to make these risks worth taking. A small win in a risky environment? For white men, who feel the world is a safe place, a small win is worth it. But for many women, it's not.

Two Yardsticks for Measuring Risky Decisions

Are there strategies for figuring out when to take a risk? I was torn on whether to include this advice because I don't want to send the message that women are, underneath it all, fearful of risk. The data doesn't support that. But whether you're a man or a woman, someone who takes too many risks or too few, you need strategies for evaluating new opportunities. There are any number of risky decisions you might be considering. Maybe you're thinking about taking a new job or quitting your current one. Maybe you're trying to decide whether to invest your time in a project that your friends think is a dead end but that you believe is just beginning. I'm going to offer two tools, two yardsticks you can use to measure a daring move and whether it's headed in the right direction.

The first tool is the 10-10-10 rule, developed by the journalist and author Suzy Welch. The purpose of this strategy is to help you look at a decision from three angles, with the hope that one of those vantage points will provide a pop of clarity. In her book *10-10-10,* Welch offers three easy-to-remember questions: "What are the consequences of this decision in 10 minutes? In 10 months? In 10 years?"[75] Simple, yes, but potentially quite powerful. The goal isn't to constrain you to those exact numbers—you could think about two days, six months, and seven years from now. The goal is for you to think about the immediate consequences, the impact your decision will have in the foreseeable and imaginable future and in a distant part of your life, a time far enough in the future that you can't predict the intervening details or events but you still have clear hopes for yourself. Imagining forty years out is probably too far. The idea is that all too often when we're trying to make a decision, we're focused on one, maybe two of these time frames, but wisdom might lie in considering all three.

The second tool for sizing up a risk is something called a premortem, a strategy discussed in the bestseller *Thinking, Fast and Slow,* by Daniel Kahneman, a Princeton University professor and Nobel Prize–winning economist who has been studying reasoning and decision-making for over forty-five years. You may be familiar with a postmortem, which is what you do when a project or event is over, but a premortem is just what its name suggests—a step you take *before* the project launches, before you've committed to a plan of action and the risks that come with it. The concept is simple. Once you have a concrete plan on the table, bring together the key people who know about the decision you're making and say, "Imagine that it's a year into the future and we've gone ahead with our current plan. The result was a disaster. Take five to ten minutes and write down a brief play-by-play of that disaster."[76]

You might not be immediately impressed with this strategy. You're thinking, *But I've already asked "What could go wrong?" a dozen times.*

But that question involves looking forward, to possible events in the future, whereas the premortem involves looking back. (A premortem is similar to the look-back we discussed in chapter 1.) Looking back may not seem like much of a shift, especially since it's all in your imagination, but this small shift in perspective can be profound.

Consider these two questions: "How likely is it that an Asian American will be elected president of the United States in 2024? Why might this happen? List all the reasons that come to mind."

Before you read on, take a moment to think about this future possibility and generate some ideas.

That was looking forward. Now consider these two questions: "It's 2024 and an Asian American has just been elected president of the United States. Why did this happen? What events might have preceded this one? List everything that comes to mind."

If you're like most people who've been asked these questions, a wider variety of vivid details come to mind in the second, hindsight scenario. It's not just that you're getting a second chance to think about the same event — people generate better answers to the hindsight questions even if they never heard the first ones. Deborah Mitchell at the University of Pennsylvania, J. Edward Russo at Cornell University, and Nancy Pennington at the University of Colorado collaborated on a project and found that people who are given the second, hindsight scenario generate 25 percent more reasons than people given the first, foresight scenario.[77] Perhaps even more important, people generated more specific and concrete reasons in the hindsight scenario. When we think about future events, we're content to think in broad generalities, but when we think about something that has already happened, we feel a need to provide more convincing explanations. This is why the premortem is so effective — it's looking back at a fictional event as though it's happened. You've always heard that hindsight is better than foresight, and that, remarkably, includes imaginary hindsight.

Know Your Anxiety

Sometimes we hesitate to take risks because we feel anxious. Imagine that you're sitting in a morning meeting and there's a discussion about purchasing some incredibly expensive software and most people are nodding along. You're thinking, *I can't believe we'd consider this, given this year's budget. Isn't anyone going to point out why it's a bad idea? Should I say something?* As you weigh whether or not to speak up, you notice you have butterflies in your stomach. Your palms are damp. As we saw in Susan Fisk's research, women often feel more anxiety than men do when they're in a risky setting at work. This uneasiness seems like an important cue that speaking up right now is a bad risk, one you shouldn't take. And once you notice you're anxious, your mind can fill in plenty of reasons to justify why you should stay quiet — *I'm too new and I'm going to embarrass myself* or *I'm too old and I just don't understand the technology* or *If everyone else in the room is fine with this, I must be missing something.*

Here's something to try the next time this happens: Ask yourself if there might be something else making you anxious, something outside of this moment. Is there anything happening today or this week, even something completely unrelated, that might cause those butterflies? You might be anxious because you're having dinner with your ex later that night or because you're giving a presentation that afternoon. When you sat down for this meeting, maybe the person next to you leaned over and whispered, "Did you get the e-mail about the office move?" Once anxiety is triggered, it seeps and lingers, and it doesn't come with a handy label that reads *By the way, I'm anxiety about dealing with your ex.* The bad feeling in the pit of your stomach is the same no matter what caused it.

Researchers have found that when people feel anxious about a threatening upcoming event, they become incredibly cautious. In one experiment, risk-taking dropped by 85 percent when people were anxious about an upcoming speech they had to give. Even risks that were completely unrelated, such as getting a flu shot, risks that people would ordinarily shrug off, seemed dicey. But the researchers found they could correct the problem simply by saying, "You may feel anxious because people often get anxious when preparing to deliver a speech." That one line offered much-needed perspective, and once the subjects were reminded of why they were actually nervous, their perception of risk returned to normal levels.[78]

When you feel anxious, it's easy to assume that the uncomfortable feeling must have been triggered by whatever is happening right that second, that you sense danger and that you've waded into risky waters. The anxiety truly could be related to the risk you're entertaining, but before you make that assumption, try reminding yourself of the other stresses and unknowns in your life that could be triggering your anxiety. You may still decide that you're not going to take the risky route, but you'll be more discerning.

Setting Tripwires

One thing that distinguishes great decision-makers from mediocre and poor ones is that great decision-makers reevaluate. Whether you've made a bold, risky choice or you've played it safe, it's judicious to take stock later, to take an honest look at a choice you made in the past and, given what you know now, decide whether you'd make it again.

Setting a tripwire is one way to ensure you reevaluate when it's most needed. A *tripwire* is a military term dating back to the early 1900s. In a war, a tripwire is a wire that runs just above the ground; when an intruder trips over it, it sets off an alarm, alerting others that an enemy or trespasser has entered an area. In decision-making, however, a tripwire prevents you from going too far down a particular path if it's proving to be the misguided one.[79] Digital tripwires are commonly used with stock investments. Let's say you're new at investing and you're looking to buy stock in Southwest Airlines. You notice the price is hovering between $33 and $34 a share, and you decide to invest $5,000 in Southwest, which feels a little risky, but you think the stock price will go up. Just in case the stock price goes down, however, you tell yourself, *If the value dips below $30 a share, then I'm selling it.* In fact, you can set up a system — a kind of tripwire — to automatically sell your stock if it falls below that price. There's a chance you might lose some money, but you've ensured that you'll adjust your investment before you lose very much, which makes it easier to take a risk.

Tripwires aren't just for investments; you can find compelling ways to use them for taking other risks. My husband and I set a tripwire when we first moved from Pittsburgh to Seattle. As I described in the introduction, he received an offer for his dream job and we had to decide if we were going to make the three-thousand-mile move before I'd had even a single interview. What made it possible for us to take that huge risk was that we agreed to a one-year tripwire. We'd move to Seattle, and if I hadn't found a fulfilling job within one year, we'd both look for new employers and move again. Knowing that we had that tripwire, that we weren't putting my career on permanent hold, made it possible to take that risk. But, thankfully, I didn't have to wait a year — I was offered a job three months after he started his.

The beauty of the tripwire is that if you set one ahead of time, you'll feel less pressure to stick with what's proving to be a poor decision. As Carol Tavris and Elliot Aronson observe in their illuminating book *Mistakes Were Made (but Not by* Me), people all too often stick with a decision they've made simply because they've made it. It's painful to recognize when we've chosen poorly, so rather than admit to a mistake, we try to justify our choices. We feel pressure to prove, even if it's just to ourselves, that we're headed in the right direction. No one wants to backtrack and look foolish. So we try to ride out that decision until the circumstances change — until the stock market bounces back up, until employers are hiring again, until we find those weapons of mass destruction — basically, until we look smart again. But if you've set a tripwire, you've planned for this contingency. You were clever enough to see that failure was a possibility, and you're reevaluating before things get worse.

This chapter began by asking what comes to mind when we think of men and women and risk. When a male takes a risk and fails, society gives him a second chance. People encourage him to climb back to the top of the fire pole and give it another try. They'll even coach him so that he doesn't make that same mistake again.

But if a female takes a risk and fails, especially in a career that's historically male, then people doubt she's really cut out for what she's trying to do. The message she hears is she should really play someplace else.

We have to think about what this unconscious bias around risk means for the leaders we choose and the leaders we keep. When a leader makes a mistake in a powerful role that seems better suited to the opposite sex, we're quick to judge and scrutinize. Remember the female police chief and the male president of a woman's college? Falter once, and people are quick to call those leaders incompetent and inclined to give those weighty decisions to someone else. If half of the powerful roles in the United States evoked the image of a woman, then the men in these leadership roles would face a formidable challenge. But few men know that challenge. Almost every powerful role in this country brings to mind the image of a man. His poor judgment we shrug off. But hers? We count it against her twice.

If we stop at this assessment, it's hard not to be depressed. So let's push our thinking a little further. Many societies see men as the real risk-takers, as the bold actors willing to try something new, insisting that risk-taking is part of what men are supposed to do. To date, however, hiring a woman to lead or promoting women to the top ranks has been something many men have been reluctant to do. Ironically, many would rather go the safe route and promote a man.

If we stop at that irony, it's hard not to be frustrated. But if we leverage the conviction that men take risks, if we appeal to that risk-taking spirit and funnel it toward taking a risk on women, we might find more women in power. I'm not saying it will be easy — it will be hard to let go of those lessons all of us learned on the playground. And I'm not saying that we should keep thinking of men as visionaries and women as wallflowers. If we think about the realities of our environments, our social cues, and what women are really doing out in the world, we might just be able to change our views about women as risk-takers and more people may be able to recognize and respect the chances women take. In the meantime, I'll take the improvement that would come from harnessing the risk-taking that men already embrace and encouraging them to look out for opportunities to acknowledge, promote, and support more women leaders.

Chapter 3 at a Glance: Hello, Risk-Taker

THINGS TO REMEMBER

1. People feel more comfortable taking a chance on a man than on a woman.
   - Venture capitalists are more likely to invest in a proposal when it's pitched by a man than by a woman.
2. Risk-taking is not a personality trait. It's a skill.
3. Don't forget the lessons from the fireman's pole. Parents encourage their sons to practice taking risks on the playground,

but they find it hard to let their little girls be anything but safe.

4. When a risky decision fails, it's more tolerable if it was made by a leader in a gender-appropriate role. Women who take risks and fail in traditionally male roles are penalized more than men in those roles with the same poor judgment.

5. Women take more social risks than men, sometimes by necessity.
   - Examples: Speaking up in a meeting full of men; single motherhood

6. When people are just learning a skill, men take more risks than women, but once they become professionals, the differences in risk-taking typically disappear.

7. Precarious manhood is a problem. Cues in the environment that threaten masculinity prompt men to take more risks than they ordinarily would.

8. Beware the white-male effect. White men find the world safer than other people do and that means a behavior that seems risky to women (or African American men or Hispanic men) feels safer to white men.

THINGS TO DO

1. When talking about your successes, draw attention to the successful risks you've taken too.

2. Use the 10-10-10 strategy and the premortem as tools to gauge whether the risks you're considering are smart calls.

3. If you feel anxious about a risky decision you're considering, ask yourself if anything else might be making you feel that anxiety, putting you into anxiety mode. Often this is something unrelated, but the mood is infectious.

4. Set a tripwire. Having a concrete plan to reevaluate makes it easier to take risks.

4

Women's Confidence Advantage

ON THE FIRST DAY of classes at Harvard Business School, all of the men and women settling into their seats were equally qualified to be there—college valedictorians of both sexes, everyone with near-perfect test scores and diverse and impressive résumés. But after classes began, the equal standing started to ebb away. Men earned higher grades. Men dominated classroom discussions. In an article that made the front page of the *New York Times,* Jodi Kantor noted how many women "sat frozen or spoke tentatively."[1] Harvard's gender gap became painfully apparent as commencement and the announcement of the Baker Scholars approached. If you're at Harvard Business School, you want to be a Baker Scholar, a distinction that indicates you're in the top 5 percent of your graduating class. It's like an Academy Award nomination—it opens doors professionally and personally—but it has been disproportionately opening doors for men. As recently as 2010, women made up almost half of the students in the graduating class, yet they were only one-fifth of the Baker Scholars.[2] Had this inequity happened once, Harvard could have it shrugged off, perhaps blamed it on a sexist incoming class or a new curriculum, but this had been the pattern for over a decade.

Harvard recognized that something was wrong with the system; something was either favoring men or handicapping women. When the university hired its first female president, Drew Gilpin Faust, in 2010, she immediately set about making sweeping changes. She appointed new administrators; the business school hired more female faculty; and professors rewrote classic case studies to feature more female protagonists. As Robin Ely, a senior associate dean at Harvard Business School, put it: "We treated the gender gap as the canary in the coal mine—a signal that our culture may have been more supportive of some students than others."[3] In 2014, the dean of the business school, Nitin Nohria, even publicly apologized to women for how Harvard had treated them. "The school owed you better," he said to HBS alums in San Francisco, "and I promise, it will be better."[4] Feminists applauded Harvard for trying to create a warmer climate for women.

But the business school also did something that will strike some as patronizing. They gave lessons in hand-raising.

"Reach up assertively!" the workshop leader said, demonstrating with her own arm rod-straight in the air.[5] She was a second-year student who was coaching the incoming class as part of their class-participation workshops. "No apologetic little half waves!" she admonished.

Were women the only ones invited to these coaching sessions? I interviewed three women who had graduated from Harvard's MBA programs across two different cohorts. "The workshops were mandatory for both men and women when I went through the program," recalled Mia, who went to the class-participation workshop in 2011. But by 2013, the workshops had become optional. Once optional, few men attended. One student, Alice, remembered seeing only women in the room in the fall of 2013. "The men were definitely invited," she said. "The Women's Student Association runs that workshop at the start of the year and sends out the invitation to all students. Anyone can attend. But you have to understand, there are so many events to choose from." The implication is that women self-select to attend and men don't. On the one hand, having the Women's Student Association sponsor the event ensures that the women in the MBA program can tailor the workshops to their priorities. On the other hand, if a women's group runs the event and sends out the invitations, most men won't see it as targeted to their needs.

One of the themes in the workshops centered on projecting confidence: When a student had an opinion or wanted to answer a question, she was to ignore all doubts and commit. These were not, of course, timid eighteen-year-olds who were offering their opinions for the first time outside of the family dinner table. Many of these women had graduated with top honors from competitive colleges and universities. They were already successful investment bankers, advisers to chief executives, and international consultants. Women who had held their own in boardrooms were asked to hold up their hands on cue.

Was the class-participation program successful? Harvard's Business School made multiple changes simultaneously, so it's not possible to isolate the impact of the hand-raising seminars. But women's contributions in class did increase. In-class participation makes up a whopping 50 percent of a student's grade in many courses at Harvard Business School, and once men and women were participating on a more equal footing, the grade discrepancy between them disappeared. And women weren't performing well only in the middle of the pack; they held their own at the highest levels. In 2013, even though women were only 35 percent of the graduating class, they made up 38 percent of the Baker Scholars, the highest percentage in Harvard's history.[6]

What should we learn from this story? Is the takeaway message from the HBS example that women should *always* be as confident as men? That women should build, exude, and, if necessary, feign confidence in order to succeed? Some will hear about Harvard's successful hand-raising program and draw that conclusion. And several popular books and magazines articles have reinforced that message, telling women they need to dial up their confidence and keep it up, way up, if they want to achieve like men.

I'm going to argue for a different takeaway. Sure, in some instances, confidence is key. If you're trying to stand out in a room of ninety students, then a high degree of confidence gets you noticed. But if women want to have a meaningful and lasting impact on their organizations, they should also be cognizant of when confidence is an asset and when it's a liability. As we'll see in this chapter, big confidence often leads to shortsighted choices. If you want to be a perceptive decision-maker, you need to know when to dial your confidence up and when to dial it down. Do women have more modest levels of confidence than men? In some cases, they do. But I'll show that if you're one of those women with more modest confidence levels, there's an excellent chance you hold the key when crucial decisions are being made.

Self-Concept? Meet Abilities

Psychologists often discuss the difference between confidence and overconfidence. If you're confident, there's a realistic alignment between your actual skills and abilities and how you perceive those skills and abilities. Put simply, confidence is having an accurate sense of what you know and what you can do.[7] Let's take something measurable. You've decided to sign up for a 5K with a friend who is an experienced runner, but it's the first 5K you've ever done. Perhaps you think you could run a ten-minute mile, so you hope to finish a 5K (which is 3.1 miles) in a little over a half hour. You line up with the other people who plan to run a ten-minute mile, waving to your friend as she lines up with the faster runners, and sure enough, you happily cross the finish line at thirty-one minutes. That's an appropriate level of confidence, or what some scientists call well-calibrated confidence. Your abilities line up with your beliefs.

Overconfidence, by contrast, is the all-too-human trait of overestimating one's own skills, personal qualities, and knowledge.[8] If you're overconfident, your beliefs exceed your abilities. You think you can run a seven- or eight-minute mile even though you haven't run a mile since college. Why do you believe that you can run that fast? Perhaps because your friend can, perhaps because you see people who look older or less fit than you lining up at that pace, or perhaps simply because you're thinking, *Eight minutes seems like a really long time.* You join your friend and the other runners who plan to do eight-minute miles and when the race starts, you push yourself, hard. Halfway through the first mile, you have to stop because you can't catch your breath. You have a terrible stitch in your side, and you have to walk for three minutes before the pain in your knee disappears. Overconfidence is when someone thinks, *I'm not sure how good other people are at this, but I can't believe that that many people are better than me.* It happens all the time.

People can also be underconfident and underestimate their true abilities. At first blush, it might sound as though underconfidence means you're delightfully surprised when you perform better than you expected. Sometimes that happens and people exceed their abilities. But all too often, underconfidence undercuts performance. When people underestimate their abilities and then encounter a problem or a challenge, they give up quickly. It's as though they say to themselves, *I can't do it, so why try? Let's move on to something I can do.* What about our 5K runners? The underconfident ones join the slow-moving walkers at the back of the line or, in many cases, don't show up on race day at all.

If You're Guessing and You Know It,
Raise Your Hand

Are most of us highly accurate in assessing our abilities? No. Men and women both tend to be overconfident in two areas. First, most of us believe that we'll be relatively skilled at anything that sounds easy, familiar, or mindless. Ask someone how long it takes him to tie his shoes and he'll err on the fast side — after all, that's something even kids can do.[9] People believe it will be straightforward to draw a bicycle (it's much harder than you think) or that it's easy to eat five saltine crackers in under a minute (without a drink of water, the average person can swallow only two).[10]

Second, both sexes tend to overestimate how well endowed they are with the qualities that are deemed by society as important. Researchers call this pattern the better-than-average effect, or, more affectionately, the Lake Wobegon effect, after Garrison Keillor's fictional town where "all the women are strong, all the men are good-looking, and all the children are above average."[11] Most of us overestimate how well we listen, how smart we are, and how kind we are to strangers.

So most of us tend to think we're just a little better, a little faster, and a little more impressive than we truly are. But is one sex more accurately calibrated than the other? Are men or women generally more accurate at estimating their abilities? Mary Lundeberg, now an educational psychologist at Michigan State University, was one of the first researchers to ask this question.[12] In the early 1990s, Lundeberg and her colleagues looked at how undergraduate and graduate students performed on their final exams across several psychology courses. Each question on these exams had two parts: a regular question about the course material that counted toward the student's grade and a place for students to rate their confidence in the answer they'd just given. Students could indicate anything from *Very certain* to *Pure guess* for each answer. The instructors explained that the confidence rating would have no impact on scores — whether the students were certain or guessing wouldn't change their grades on the exam or in the course.

The findings were revealing. Men and women were equally and appropriately confident about their performance when they chose the correct answer. When people were correct, they knew it and said so. But a different picture emerged when students made mistakes. Women tended to be appropriately hesitant about the questions they got wrong — they knew they were guessing on those questions and were willing to admit it. Men, however, still indicated high confidence in their incorrect answers and marked *Very certain* or *Certain* when they were wrong more often than women.

What's happening here? It could be that men couldn't tell when they were guessing, or perhaps they knew when they were guessing but were still highly confident that their guesses were correct. This difference may reveal only what the two sexes were willing to admit. Women might have been more willing to admit uncertainty, whereas men might have believed that uncertainty would hurt them, that admitting anything less than certainty would hurt their score, their good standing with the instructor, or their self-image. Whatever the underlying motive might be, the way women expressed their confidence ebbed and flowed, making their confidence more closely tied with what they actually knew. Men's expression of confidence just flowed, which meant their estimates of performance didn't reflect how little they knew.

In the years since Lundeberg and her team published their work, many researchers have found that men outshine women in overconfidence. Both men and women tend to think they are better than average, but it's usually men who think they are much, much better than average.[13] Take intelligence. If you approached a random crowd of people on the street, you'd expect that about 50 percent of the group would be of average intelligence, about 25 percent of the people would be above average, and about 25 percent would be below.[14] But if you took a poll, few people would look around and see themselves as intellectually mediocre. If you took the poll in America, about 71 percent of the men would say that they were above average in intelligence while 57 percent of the women would make that same claim.[15] This basic pattern of what we could call male hubris or female humility around intelligence has been replicated in Africa, Europe, the Middle East, and East Asia, so it's not limited to a particular language, educational system, or culture.[16]

Can we predict in what areas men and women will differ in their confidence? Several researchers have deduced that women tend to be appropriately confident on activities that are deemed "feminine." For instance, women are typically accurate at predicting whether they have relatively high or low emotional IQ. Emotional IQ is conventionally thought to be an area where women have an advantage, and here, women are good at predicting their performance. Women who have high emotional IQs know it and women with low emotional IQs recognize they find it hard to gauge what others are feeling.[17] Likewise, women tend to be good at guessing how much trivia they'll know about films and TV shows aimed at female audiences. Think of TV shows like *Sex and the City* or *Downton Abbey,* or movies such as *The Help* or *Wild.* A woman might shrug and say she doesn't know much about these shows, but that's exactly the point. The research predicts that if you're a woman, you're good at estimating how much you do or don't know about these shows. Your beliefs about your knowledge line up with your actual knowledge.

Women tend to be underconfident about their knowledge and abilities in areas of the world that are deemed "masculine." Most women predict that they know less sports trivia than they actually do and assume that it will take them longer to figure out a new computer than it actually does, both of which are areas where society tells us men tend to excel.[18] Even women professionals who make their living in male-dominated fields tend to underestimate how much they know. One study found that women who were highly knowledgeable in the world of finance, another field traditionally associated with men, were underconfident of their abilities and predicted that they would know less about finance than they actually did. The highly knowledgeable men, however, knew it and said so.[19]

Here's the interesting twist: Men don't have a parallel underconfidence for tasks that are considered feminine. If anything, research suggests male overconfidence is undeterred by knowledge or skills that are deemed feminine. In one study, men predicted their scores on a quiz about women's TV shows and films, and they presumed they'd know much more than they actually did. In another, men were overconfident in their own emotional intelligence. In this fascinating research study, men and women first estimated their emotional intelligence and then took tests to measure their actual ability to recognize, understand, and manage emotions. Men estimated their emotional IQ scores would be higher than women's, but the reality was the reverse.[20] Women were significantly better than men at solving emotionally laden problems, such as understanding how one feeling changes over time into another. Were the men humbled after taking the test? Did they reassess their abilities and dial down their confidence? Not really. They still gave themselves higher marks than women did.

Are men more accurate about clichéd male interests? Again, not really. Research shows that men are either appropriately confident or overconfident in "guy things." Men have an accurate sense of how much sports trivia they'll know, but they tend to be overconfident in how well they trade stocks, how safely they drive at night, and how many rounds of roulette they'll win, even though it's a game of chance where they can't control the outcome.[21]

Do all studies find that men enjoy a bigger slice of overconfidence than women? No. When researchers invent a new task that isn't seen as either masculine or feminine, they typically find that men and women both enjoy overconfidence. One team of psychologists asked residents of Pittsburgh, Pennsylvania, to estimate what the high temperature would be in that city on randomly selected dates.[22] As we saw earlier, familiarity gives us a false sense of expertise, and both male and female Pittsburgh residents thought they were better forecasters than they actually were. Guessing the temperature didn't have any gender baggage or expectation.

So what's the big deal? So what if women think they'll have trouble answering questions about sports? So what if men have no sense of how much they know about soap operas? The cause for concern lies in the fact that large swaths of American culture are roped off as "better suited to men" and that women take these barriers to heart, underestimating their abilities when they cross into those male-dominated zones. In the workplace, many tasks are considered masculine. Men are expected to be better at making managerial decisions, negotiating, coming up with financial projections, and communicating directly and succinctly, which suggests that high levels of confidence in the professional workplace are the norm for men, but not women.[23] That's an awfully big part of a working woman's life where she is likely to incur underconfidence and therefore undercut her own success.

Do men actually see themselves as better leaders than other people see them? Samantha Paustian-Underdahl, a management professor at Florida International University, knew that there was a puzzling contradiction in the leadership literature; roughly half of the studies reported that men make better leaders, while the other half said women do. How could both be possible? It can make the skeptical types throw up their hands and say that research on gender and leadership is meaningless. But Paustian-Underdahl wondered if the difference might lie in who did the evaluations. She led a team that took a closer look at ninety-nine research studies on leadership skills among men and women.[24] When employees were asked to rate their *own* abilities as leaders, men gave themselves higher ratings and women gave themselves lower ratings. For those studies, men looked like the more competent leaders. But when the study also included other people's ratings, those numbers didn't line up with the self-assessments. Work colleagues, subordinates, and supervisors more often said women, not men, made better leaders. Paustian-Underdahl didn't examine whether it made a difference if the colleagues were male or female, but her findings suggest that if your organization does annual performance reviews, men's self-evaluations tell one story and women's self-evaluations tell another.

These findings fit the patterns of overconfidence and underconfidence we see in men and women. For decades, the assumption has been that men make better leaders than women.[25] Social scientists even have a catch phrase for this belief: *manager = male.* The perception that men make better leaders has been changing in recent years, but Paustian-Underdahl's analysis suggests that men still believe they're a bit better at leading, particularly at the executive levels, than they actually are, whereas women underestimate how well they can and do perform.

Many scientific, government, and professional roles are traditionally associated with men, and we can imagine the benefits that men enjoy from the overconfidence that results. If a man thinks, *I can lead better than two-thirds of the people in here,* then he's probably going to ask for a raise (and make a strong case for getting it). If he's convinced his résumé doesn't reflect his simmering potential, he's going to seek out jobs that are harder than the one he has now. And we aren't just imagining these benefits; men do ask for raises and seek promotions more often than women. According to Linda Babcock, an organizational psychologist at Carnegie Mellon University, for every one woman in an organization who asks for more money, there are four men who do.[26] And a 2008 report from Hewlett-Packard found that women applied for jobs within the company only when they believed they met 100 percent of the posted qualifications. Men, by contrast, applied when they met only 60 percent of the job requirements.[27]

If the conversation ended here, if this were all the data we needed to see, then the conclusion would be clear — women should brim with confidence. Matching men's high levels of self-confidence would be women's logical goal in order to succeed. That builds on the premise of nearly every leadership book: if you want to be successful, do what the most successful people are doing. But now we need to turn to decision-making. What are the most successful decision-makers doing? What roles do confidence and overconfidence play when it's time to make a big decision, one that may determine your future success?

Ignoring the Two O'Clock Rule

Friday, May 10, 1996. Mount Everest. It was a busy day on the highest mountain in the world, with twenty-three climbers reaching the summit that afternoon. These climbers had trained for months, some for years. Despite all that preparation, May 10 would be the deadliest day in the mountain's history. Five of the climbers didn't make it back down alive, including the two men leading the teams. Not only did 22 percent of the group die on this trip, a terrible track record for any expedition, but several other climbers only barely survived. They spent hours stumbling and wandering in the dark, brittle and blue with frostbite.

You might be familiar with the story that's compellingly captured by Jon Krakauer in his popular bestseller *Into Thin Air*. It's tragic, but that's not the only reason it captures our attention. This story is also strange. The climbers had reached the top of Mount Everest, and their bodies had made the difficult adjustments to the extreme altitude and the low oxygen levels, so it seemed like the hardest part of the journey was behind them. It would be like a group going on a rafting trip and making it safely through the most technically challenging rapids only to have one out of five people drown in the last fifteen minutes. Many have tried to understand how things went so terribly wrong after the group's substantial achievement. Part of the problem was obvious — a sudden storm blew in and temperatures plummeted to forty degrees below zero — but storms are to be expected on Mount Everest, and the leaders, who were highly experienced on the mountain, had seen plenty. A newer, post-Krakauer analysis reveals that overconfidence was rampant. Prior to the climb, Scott Fischer, one of the two most experienced leaders, told his team, "We've got the Big E figured out . . . These days, I'm telling you, we've built a yellow brick road." He even told reporters, "I'm going to make all the right choices." The other leader, Rob Hall, also showed his overconfidence, bragging that he "could get almost any reasonably fit person to the summit." When Krakauer, who went on this expedition as a journalist, expressed doubts about reaching the summit, Hall said, "It's worked 39 times so far, pal."[28]

The leaders were convinced they had the mountain figured out, and their overconfidence in their abilities seems to have led to poor judgment on the last day of the trek. The two o'clock rule is a good example. Both Fischer and Hall had talked about a strict deadline for reaching the mountain summit. Fischer often admonished his team, "Darkness is not your friend," so any climber who hadn't reached the top by one or two in the afternoon was supposed to turn around, no matter how close he or she was to the goal. But when the two o'clock deadline came and went, the leaders didn't enforce that rule. One of the climbers didn't reach the summit until four, and Hall gave him that leeway. Neither that climber nor Hall made it back down the mountain.

Overconfidence was not the only obstacle that day—oxygen levels were especially low, which slowed people's expected climb and descent rates.[29] But researchers who have looked closely at the accounts know that overconfidence made the team leaders too sure of themselves, too willing to take risks they had always advised other people not to take.

Studies centered on disastrous leadership decisions, mistakes that no one wants to make again, are finding that situations where terrible decisions are made commonly have overconfident leaders, decision-makers who overlooked available data because they believed they knew better. Overconfidence in decision-making has been blamed for the global financial crisis of 2009, for the British Petroleum oil spill in the Gulf of Mexico in 2010, and for Japan's nuclear power plant crisis in 2011.[30] That's one disaster a year for three years, all of which could have been averted or mitigated if there had been more conscientious debate at the top. But overconfidence squelches debate. It sends the would-be debaters to another part of the building. In the case of the global financial crisis, several people crunching possible doomsday numbers spoke up, arguing that backing subprime mortgages was far too risky. In 2006, a full two years before the stock-market crash, a senior vice president of Citigroup, Richard Bowen, began warning the company's board of directors that 60 percent of their mortgages were problematic.[31] He sent weekly reports. Eventually, the other executives and members of the advisory board came to consider Bowen such an annoyance that they relieved him of most responsibilities and notified him that his physical presence was no longer required at board meetings.[32] When you're overconfident, as Citigroup execs were, alarming information doesn't alarm you. The data (and the bearer of the data) must be wrong.

Overconfidence didn't single-handedly cause any of these disasters.

Greed, a reluctance to impose standards, and the priority of private stakeholders over public good all played a role. In the case of Japan's nuclear power plant, there was an earthquake followed by a tsunami, so human error was only part of the equation. But overconfidence is a multiplier. Overconfidence increases the chances that each hypothetical problem will become a real problem because it gives leaders a reason not to seek more data or fully examine the data in front of them. When you're convinced you know better than most people, you don't enforce even your own policies, like the two o'clock rule.

You might be thinking, *Well, I'm not leading mountain expeditions and I don't oversee a nuclear power plant, so if I'm overconfident in my decisions, no one gets hurt.* But overconfidence leads to costly choices even when those decisions aren't disastrous. Two researchers at Columbia University, Mathew Hayward and Donald Hambrick, tracked fifty-three firms that were each involved in a large merger or acquisition of another company. Hayward and Hambrick wondered whether a CEO's sense of self-importance might interfere with his judgment when he bought another company. But the two men didn't give CEOs a paper-and-pencil test to gauge overconfidence. They simply picked up the newspaper. Hayward and Hambrick searched through major newspapers and business magazines in the three years leading up to each acquisition and identified the CEOs who were each extolled as the ticket to a company's success. For each favorable article a CEO received, he paid an average of 4.8 percent more for a company he acquired than did a CEO who hadn't been publicly put on a pedestal. If two favorable articles were written, a CEO paid an average of 9.6 percent more. A 9.6 percent premium might not seem like much, but if an acquisition costs $100 million, that's an extra $9.6 million dished out in a single buy, an amount that even those of us without accounting degrees would find pretty high.[33]

Why would appearing on the cover of *Bloomberg Businessweek* be bad for business? Hayward and Hambrick reasoned that hubris was to blame and that overconfidence clouded the executive's judgment. So CEOs who are lauded for their work actually cost their companies more and are less effective at this crucial decision-making piece of the job, and those who are less visible, less praised, are potentially better at assessing market value. Overconfidence handicaps good judgment when things are going well — perhaps *especially* when things are going well.

What Confidence Actually Signals

What do these lessons in overconfidence suggest? Leaders who aren't sure that they have all the answers, who have more accurate self-assessments than their overconfident peers, are going to keep scanning their environment and turning to the data as they survey their options, and women are more likely to fit that description. With their more reality-based levels of confidence, women tend to go into decision-making with their eyes open. That's women's confidence advantage as decision-makers. If a woman feels humble compared to the men around her when a decision is being made, she's probably one of the most valuable assets in the room.

You might be thinking, *Well, men can master their confidence levels too,* and that's certainly true — should they decide there is cause to do so. In the meantime, if women are going to utilize this advantage, they have to see confidence as the right kind of tool. Too many people believe that confidence is a hammer and every problem is a nail. By now you've seen that's not the case, certainly not for decision-making. Confidence is more like a volume knob. You can turn your confidence up when you need to be heard, and you can turn it down when you need to listen and make a hard decision. And in this way, people have been hearing the wrong message about what women need to do. Because women err on the side of reasoned confidence, they are better equipped to learn how to manage their confidence levels. I suspect it's easier for the timid to learn how to dial things up than it is for the bullish to learn how to dial things down.

Why do people see high confidence as a tool for every occasion, not a dial that can and should be adjusted? Most of us misinterpret what confidence signals. We'd like to think that high confidence indicates that a person is right. If someone appears confident, that's a sign that the person knows what he or she is talking about; if you feel confident yourself, that's a sign you know what's coming. When you experience a surge of certainty, you feel like you've gotten a glimpse of the future and that future is good. (Most of us don't say, or even consciously think, *I've seen the future,* because that sounds crazy, but it is the unspoken promise of certainty.)

But being confident isn't the same as being right. This seems obvious when it's stated plainly, and research confirms it. We can turn to the criminal justice system to see how confidence is a terrible proxy for accuracy. Studies show that detectives and police officers who are sure they always know when someone is lying aren't actually much better at figuring out if the guy across the table is telling the truth than the average person.[34] Eyewitnesses to a crime who swear that they can positively identify a criminal pick the wrong person from a lineup almost as often as those bystanders who shrug and say they'll do their best.[35] In their illuminating book *The Invisible Gorilla,* Christopher Chabris and Daniel Simons call the tendency to confuse confidence with correctness an "illusion of confidence." You can be certain of your judgment and still be wrong.

You might think, *But someone who happens to witness a crime is very different from a trained expert. Professionals know when to dial back their confidence.* That reasoning seems valid, but let's take a closer look. Consider television pundits, who make a living "commenting or offering advice on political and economical trends." Political scientist Philip Tetlock from the University of Pennsylvania has found that those television pundits who express the greatest confidence in their predictions are the least accurate.[36] If a commentator predicts that a particular candidate is going to be elected and the pundit turns out to be right, he takes that as feedback that his political forecasting skills are strong. His confidence goes up a notch. But what if the opposing candidate wins? The pundit doesn't have to see that as a reflection of his predictive prowess. After all, the outcome of the election could be explained by all kinds of extraneous factors, such as low voter turnout or a comment by a politician in a different state that deeply offended undecided voters. Who could have predicted these things? So when the other candidate wins, a pundit doesn't have to dial his confidence down. It can stay right where it is. Always going up, rarely moving down.

If confidence doesn't signal that you're right, what does it signal? Confidence doesn't indicate what you know; it indicates that you've been able to tell yourself a good story. Daniel Kahneman explains what confidence really means in *Thinking, Fast and Slow:* "The confidence that people experience is determined by the coherence of the story they manage to construct from the available information. It is the consistency of the information that matters for a good story, not its completeness. Indeed, you will often find that knowing little makes it easier to fit everything you know into a coherent pattern."[37]

This is why the most confident television pundits are the least accurate — they've looked at fewer pieces of data. They are certain that so-and-so will win the election because they've focused on A, B, and C and ignored Y and Z. As long as you can tell yourself a good, coherent story, you don't need to look any further; all the pieces fall into place.

Knowing When to Turn That Dial

Even if we have all of the tools and skills we need, there's still the question of when to do what. How can we tell whether confidence is going to serve us well or get in the way? How do we know *when* to dial it up or down? As we've seen, a high level of confidence obstructs good decision-making but helps persuade others that you're right. If you're giving a presentation at work or you're trying to get a professor to call on you in class, you are engaged in persuasive behavior. But if you're too certain of yourself too soon, you compromise your judgment.

So how do you finesse your use of confidence? Successful leaders dial their confidence down when they're in the process of making a decision. With your confidence turned down, you listen carefully, remain receptive, and take in all of the necessary information. "Stay curious" is your mantra. But once you have made a decision, when you are pushing forward an agenda, strategically dial your confidence back up.

What does this look like in practice? Take Lila, a pulmonary specialist at one of the top-rated hospitals in the United States. She works in the intensive care unit, but she told me a story about a patient in a different wing of the hospital. This patient had undergone heart surgery and although his heart was recovering on schedule, his lungs weren't. In the days following his operation, a breathing setback meant he had to be placed on a ventilator. When the patient was still struggling to breathe, his doctor had made an incision in the front of the man's neck and inserted a tube in his trachea. Would this become the permanent solution? The heart surgeon asked if someone from the pulmonary department could swing by to offer input. Lila got the call.

She went to this patient's room and started with his chart, but, like any good doctor, she couldn't decide what was going on from that alone. Usually she'd talk with the patient, but he couldn't speak to her because he was on the ventilator. Lila could understand his mouthed words, but she needed details. Thankfully, the patient's wife was at his bedside. Lila learned the man had never been officially diagnosed with asthma or bronchial problems prior to the surgery, but, his wife told her, he had occasionally felt short of breath if he pushed himself too hard exercising. As Lila listened, she picked up the man's hand and held it to reassure him that she was paying attention to him as well. This might have been a clear-cut case, but Lila wasn't sure, so she kept asking questions. She studied the breathing machine. She asked to speak to his nurse, who was in the next bed space. "No one likes to have an invader come in to see their patients, and even though I'm wearing the right coat and have the right badge, I'm not known in this part of the hospital like I am in mine," Lila told me. "So when the nurse comes over and I ask how he's doing, the nurse says, 'Well, you can see him right there.' I said, 'I appreciate that, but I work in a unit where the nurses and doctors work closely together and I'd like a little more information from you.'" And again, the nurse said, "Well, what you see is right there," nodding first to the chart Lila was holding and then to the bed where the patient lay. Lila saw she had two problems — she needed more subjective information to judge the best course of treatment, and she needed it from this person who was disinclined to help. She kept her confidence level dialed down, both internally and externally. (I couldn't help noticing she'd kept her anger down too.) She could have peacocked her credentials, but she knew that this nurse, who had been overseeing this patient for more than a week, probably had concrete observations, and Lila needed to hear them.

Slowly, with Lila's gentle encouragement and persistence, the nurse offered more detailed information about the patient's daily symptoms, nuances of how he was or wasn't reacting to different medications. The pieces began to fall into place. The patient's heart regimen was appropriate, but there was a less invasive solution to his breathing problem. Lila reached her decision — he needed to be moved to a different unit where they could adjust his breathing regimen. He might be on a ventilator indefinitely if they continued this course of treatment. Now Lila ramped up her confidence. She called the doctor who'd requested a pulmonary specialist and said: "I would love to bring him over to our unit. We can get him off that machine. Let us take care of this for you." The patient was moved the next day.

Several months later, Lila received a letter from the patient's wife. She wrote that after sixty-six days in the hospital, her husband was finally back home. The tube in his trachea had been removed and he didn't need an oxygen tank. He was breathing completely on his own. *We were lucky you were there,* she wrote, *to get us from one part of the hospital to another. You rescued us.*

There are many tools that Lila employed to make this judgment and get this patient moved, from her knowledge of different ventilators to her interpersonal sensitivity. But let's look at how her ability to turn her confidence up or down at the right time was indispensable.

What might have happened if she had gone into that patient's room overly confident? As we learned earlier, overconfidence gives people a reason not to seek extra data. As soon as all of the information in front of them fits into a clear, tidy story, they stop asking questions. If you're expecting to feel confident, you don't want a messy story where you walk away looking and feeling unsure. So if that had been Lila, she would have read the patient's chart and found that everything there told a coherent story that fit the other doctor's diagnosis—it was a textbook case of respiratory failure—and she would have agreed that the trachea tube was the right course of action. That's one standard practice. She would have concurred that they needed to address the man's symptoms one at a time, treating the most severe symptom first, then moving on to the next concern, and so forth.

But Lila didn't settle for that story. She stayed curious, even when she had an acceptable explanation for all the obvious data, and she kept asking questions to see if each new piece of information would still fit. She wasn't afraid to discover something she didn't understand. She fiddled with the ventilator. She asked which positions in bed made the patient cough and when his coughs were wetter than others. Now she was uncovering data that didn't fit that original story. And because she kept seeking data, because she wasn't trying to experience an immediate click of confidence, it gradually came to her that all of his symptoms, even the small ones, needed to be treated simultaneously, not one at a time.

It's a small consolation, but the patient's doctor was following a well-recognized course of treatment for lung complications. Remember, Lila works at an excellent hospital, one you would probably know if you heard the name. This wasn't a problem of insufficient care; it was a problem of insufficient data and using only that data to make an insufficient judgment. Lila believes that if she'd projected certainty and superiority, that nurse who had been taking care of him wouldn't have shared those small but crucial observations that helped Lila diagnose him. But wouldn't a female nurse respond well to a female doctor? Not necessarily. It's unfair, but as we'll see soon, women and men both can be slow to warm up to women who come on strong.

So keeping her confidence turned down was important while she was still gathering data and deciding, but what if she'd continued to keep it down? What if she hadn't turned up her confidence once she decided, what if she hadn't gone to the surgeon and declared, "We can get him off that machine"? The surgeon might have kept the patient in the ICU and made the trach tube permanent, which means the patient would still be using it to breathe today. Or the doctor might have authorized the transfer even if Lila had been deferential and uncertain, but he might not have trusted Lila with all aspects of the patient's care and might have sent mixed messages to the family and the patient. If patients don't believe in their care teams, Lila told me, they don't recover as well.

What's especially impressive? Lila didn't even treat this patient. She ushered him into someone else's care, a trusted colleague who she knew would be able to help the patient breathe on his own again. She didn't size up the situation and say, *I'm better.* She sized up the situation and said, *There is better.*

Emma Stone and Wonder Woman

A survey of 509 women found that most of them (92.8 percent) agreed that confidence was a quality you could acquire over time, that it wasn't something you had to be born with.[38] But many of these women reported that a frustrating experience, like having a colleague who is overly critical or a leader who micromanages, could deflate their confidence. So what can you do in one of those moments when you need a confidence boost? Here are some tips and tricks for turning up that confidence dial.

First, lower the pitch of your voice. Don't whisper—you want people to hear you—just talk in a lower pitch. Think of the deeper voices of Emma Stone, Kathleen Turner, Lauren Bacall, Whoopi Goldberg, Rachel Maddow, and Toni Braxton. Research shows that when people talk in a lower voice, they feel a greater sense of power and confidence, and they find it easier to think more abstractly.[39] People experience boosts in their confidence and their problem-solving abilities within five minutes of starting to talk in a lower tone. It might give you just the injection of confidence you need.

Not only might you get a surge of confidence when you start speaking differently, but the people around you might perceive you differently as well. People prefer men with low voices. Men with deep, resonant voices are seen as more attractive, more powerful, and more competent than men with higher voices, and it shows up in their salaries. For every 1 percent drop in voice pitch, a male CEO makes nineteen thousand dollars more in annual compensation.[40] Recently researchers have begun asking whether we also prefer female leaders with low voices. It turns out that people expect women with higher voices to be sexier and more physically attractive, but they expect women with lower voices to make better leaders and be more competent and trustworthy.[41] Given a choice between two unfamiliar female candidates running for office, people vote for the one with the lower voice. Former British prime minister Margaret Thatcher knew this decades before the scientific world confirmed it. With the hopes of commanding more respect early in her political career, she sought out voice coaching to help her lower her vocal pitch. Some analysts postulate that it enabled her to win the 1979 election.[42]

Another potent way to give yourself a surge of confidence is to change your posture. This suggestion comes from a growing area of philosophy and psychology called embodied cognition, a cross-disciplinary field that asks how the body influences the mind. Take a quick look at how you're sitting right now. Are your arms resting on your body? Is your hand touching your face or your neck? Are your legs crossed demurely at the ankles or perhaps even tucked beneath you? Although I have to admit that all of these sound comfortable and natural to me, these would be called low-power poses. As Amy Cuddy explains in *Presence*, in a low-power pose, your body takes up less space and your arms and legs are relatively close to your body. Picture Woody Allen in almost any movie, and you know what a low-power pose looks like.

Now try a high-power pose. Lean back and put your hands behind your head, elbows out. Prop your feet up on the table in front of you. It's the relaxed, stretched-out, "I'm large and in charge" pose that we might associate with a law partner or a business vice president thinking through a decision. In a high-power pose, people literally take up more physical space. If putting your feet on the table is out of the question, stand up, place your hands squarely on your hips, and put your feet about eighteen inches apart. The first time I tried it, I felt like I was modeling for a Wonder Woman poster.

And the superhero image isn't a bad one to have in mind. These are called high-power poses because if you adopt one for as little as two minutes, research shows you'll actually feel much more confident, powerful, and in charge.[43] Women have begun to use these techniques to improve their confidence in the workplace. Liberal political pundit Sally Kohn practices with these high-power poses before she goes into heated debates on Fox News. She steps into the hallway, adopts her Wonder Woman pose, and holds it for a few minutes. Then she strides onto the set, feeling more confident, less anxious, and fully capable.[44]

Tools to Rein in Your Confidence

What if you face the opposite problem and need to dial down your confidence? Advice on how to lower your confidence is scarce, but if you want a complete toolkit for approaching all kinds of decisions, you need these strategies as well.

You can try, like Lila, simply to make an effort to listen more, but some might find that strategy ineffective. If you're being quiet in a meeting but the thought running through your head is *I have more important places to be,* you're not really taking in new information, nor have you lowered your confidence. Instead, try doing a premortem, which we learned about in the chapter on risk-taking. When you do a premortem, you imagine that you are already at some point in the future — three months, a year, perhaps five years from now — and your decision has failed miserably. Then you write down the reasons it failed. When people ask, "What *went* wrong?," they generate, on average, 30 percent more potential problems than when they ask, "What *could go* wrong?" When you ask what went wrong, you're replacing the success story in your head with a failure story, which dials down your confidence.[45] When researchers compared a premortem to three other techniques (including writing down a list of pros and cons, which is the go-to strategy for most of us), they found the premortem reduced overconfidence to the greatest degree.[46]

You can also try adopting a low-power pose. High-power poses increase confidence, and low-power poses do the opposite. Take up less space, keep your limbs close to your body, and put your hands in your lap, on your neck, or around the tops of your arms (the way you might if you were cold and trying to warm yourself). Within a few minutes, you'll probably feel like taking fewer risks and you'll feel less in charge.

"Arrogance Is So Ugly on a Woman"

You may have noticed that when I've talked about confidence, it's largely about confidence in public arenas — how Harvard MBA students participate in class, how lowering one's voice helps project authority, how two Mount Everest expedition leaders behaved when a journalist followed them up a mountain, recording their every move. Could self-presentation be a crucial part of the difference we see in the confidence between men and women?

Over twenty-five years ago, three psychologists at Bucknell University and Williams College, Kimberly Daubman, Laurie Heatherington, and Alicia Ahn, asked that question. They wondered whether girls and women privately believed in their abilities, that while they were on the bus to school or the train to work, they actually had ambitious visions for themselves, but when they arrived at their destinations and had to tell others how good they were, they became modest. They hedged. They lowered their confidence in the presence of others.

Daubman, Heatherington, and Ahn asked freshmen college students of both genders to predict their first-semester grade point averages. Some of the students had to make their predictions out loud. The student had to look at the experimenter and say, "I'm getting all As, so I expect a 4.0" or "I hear my philosophy professor is impossible, so I'm expecting to get maybe a 2.9." That was the public condition. Other freshmen wrote down their predictions anonymously on notecards and sealed them in envelopes after being assured that what they wrote would be kept confidential. That was the private condition.[47]

What the researchers found won't surprise any woman who has gone into a meeting with a great idea and left the meeting puzzled that a man in the room somehow took credit for it. First of all, there was no difference in competence. The actual GPAs that the men and women received were, on average, the same. The women were doing just as well in their classes as men. But predictions about grades varied dramatically for both genders depending on whether they'd had to announce their grades publicly or jot them down privately. Men who wrote down their GPAs privately were quite accurate. But men who made public predictions inflated the numbers. They exaggerated their abilities. When a man had to tell another person his predicted GPA, he gave numbers that were higher than the actual grades he received, making himself look rather good in the process. For women, the exact opposite occurred. When a woman had to tell someone else how well she expected to do, she lowballed her estimate, giving numbers that were significantly below the grades she actually earned. These women were modest, underselling themselves even when the only other person in the room was an experimenter, a stranger they had just met ten minutes earlier.

Women didn't misgauge their abilities; the ones who made their predictions privately, who had the chance to discreetly evaluate what they did well and what they did poorly, were spot-on, much like the men in the private condition. They knew how skilled they were; they just needed to be asked about it in the right way. This brought up some obvious questions: Was this really a confidence problem? Or had women learned to be modest? The researchers hypothesized that women had absorbed the message that it was more feminine, and therefore more desirable, to downplay their abilities, while men had received the message that it was more masculine to upsell theirs. It mirrors what one of the female Harvard MBA students said to me: "My father always told me, arrogance is so ugly on a woman."

Researchers tried the experiment again, this time with more women and men. They got the same result.[48] In the years since, researchers have given this concept a name: self-promotion. Self-promotion is about making "one's competence visible," so when you self-promote, you don't just know your abilities, you also make those abilities known.[49] And the most disturbing sex difference around self-promotion isn't that women self-promote less often than men. That's true — women are less likely to speak of their accomplishments or show that they know more than other people in the room. But what's really disturbing is the growing evidence that women are penalized for self-promotion while men are not. When women directly draw attention to their accomplishments, they are found to be less likable, less attractive, and less hirable than men who make these same comments.[50]

Rachel, a political campaign strategist, said that she asked her boss for a promotion several years ago. She followed all the advice she'd read about how women should ask for raises and promotions — she handed him a memo documenting the ways she'd been an asset to the organization, she told him how much she loved working there, and she explained how she believed she was being underutilized in her current position. She waited for his reply. Her boss looked at the memo, said, "Well, this isn't going to happen," then literally threw the paper in the air.

Was this just a case of a terrible boss? Perhaps. But the story fits the larger pattern — people don't like women who self-promote.

Laurie Rudman, a social psychologist at Rutgers University, is an expert on how people respond to women who self-promote. She compares how people react to men and women who make confident and direct statements about their contributions (such as "I respond well in high-pressure situations") to how they react when people include a softening disclaimer (such as "I'm no expert . . ." or "I don't know if this is right, but . . ."). In what's now considered a classic study, Rudman asked research participants to watch a videotape of a job candidate talking about his or her qualifications, much as Rachel talked about her accomplishments when she went in to speak with her boss.[51] How did the viewers perceive the candidate? We might want to imagine that women, at least, were receptive to strong women, but that wasn't what happened. Both sexes were more interested in hiring the man who made strong, direct statements about his skills than the woman who used the exact same language in exactly the same tone. The only time evaluators really liked the woman who self-promoted was when they had a monetary incentive to find the quickest, most intelligent person for their team. But when the evaluators were told to find someone who would more broadly "ensure a project's success," they didn't want the woman who self-promoted, although the guy who self-promoted would work just fine. In fact, for men, self-promoting is often considered an asset, and even if it isn't, it's shrugged off or ignored. But we don't grant women the same free pass.

Why wouldn't evaluators want the woman who described her relevant accomplishments directly? Isn't that what you're supposed to do in an interview? It's not that simple. Women who mention their outstanding achievements are more likely to be judged as dominant, strident, and arrogant than men who list their successes.[52] And these adjectives lead to an assessment of women as less-than-ideal candidates for teamwork.

Public promotion isn't a no-no only when women are talking about themselves. It's also an issue when women are publicly recognized by others. Consider two of the many high points in the career of Sheryl Sandberg, the chief operating officer at Facebook. The first high point happened long before she arrived at Facebook, in 1994, while she was in the MBA program at Harvard Business School. (This was, for the record, years before the hand-raising seminars.) During the summer between her first and second year of business school, Sandberg received a letter congratulating her for having one of the highest academic averages in her class. She was named a Henry Ford Scholar, but it wasn't a public distinction. No list would be posted in class, no announcements would be made, and she decided not to tell anyone at school but her closest friend. When she saw how her classmates fawned over a male student who'd received the same scholarship and announced it, she wondered, momentarily, if she was making a mistake. But the moment passed. She kept her success and her knowledge of her abilities quiet, and in her book *Lean In,* she said that that felt right, even though she had no idea why. Perhaps she even felt an extra boost of confidence to know this honor was her secret.

Compare that to another high point of her career seventeen years later when, in the summer of 2011, *Forbes* magazine listed her as one of the five most powerful women in the world. Sandberg was already a prominent figure at Facebook, and *Forbes* put her above First Lady Michelle Obama and Indian politician Sonia Gandhi. Sandberg said she felt horrified. When people stopped her in the hallway to congratulate her, she said she "pronounced the list ridiculous." When her friends posted the link on Facebook, she asked them to take it down.[53] Rather than feeling a surge of confidence with the recognition, she had to swallow her discomfort. For many women, even public figures like

Sheryl Sandberg, private success is more comfortable than public acclaim.

How to Navigate the Tricky Balancing Act of Self-Promotion

Professional women are being urged more than ever to ask for more money, bigger challenges, and better titles. "Have confidence in yourself and you'll get what you're worth," they're told. For some women, believing in themselves and their value might be the problem. But now that we have a better understanding of what confidence is and what it isn't, we can see a deeper reason why many women feel less confident than men when they want to ask for a raise or a promotion. Let's imagine two highly competent professionals, a man and a woman, both thinking about asking for a 10 percent raise. They're considering the same core set of issues: *Is the organization making or losing money? Am I indispensable? What's the market rate for someone doing my job? Is there a comparable position for me elsewhere?* So far, they need to fit the same pieces into a coherent story, namely, a story about whether they're in a good position to negotiate.

But she has more pieces to consider: *Will I be penalized if I ask for more money? Will my boss dislike me or give me crummy assignments going forward if I make this case? Will I be called difficult or spoiled?* Jennifer Lawrence, the Academy Award–winning actress, said that last worry influenced her decision not to fight for more money when a computer hack of Sony Pictures revealed that she made less than her male costars.[54]

These concerns aren't the products of a woman's imagination. They are real facts that a woman has to work into the story she tells herself, especially if she or her female coworkers have been penalized for self-promotion in the past. She may decide that she doesn't care what her boss thinks of her, or that she'll find a new job if she's treated poorly, or that being called spoiled isn't that bad, but that's still an extra piece of the story that her male counterpart doesn't have to consider because men rarely lose likability points when they negotiate and self-promote. A man may not get the salary bump or the corner office that he wants, but for him, it truly doesn't hurt to ask. Remember that people feel confident when they can tell themselves a story where all the pieces fit. With one less piece to fit in and no penalties to consider, it's that much easier for him to be confident.

A recent study confirms that women do grapple with the "Is it okay to ask?" piece of the puzzle. One research study compared the ways nearly twenty-five hundred job applicants responded to two different job ads.[55] When a starting wage was specified and it wasn't clear whether that salary was negotiable, men were much more likely than women to initiate salary negotiations and ask for more money. When the job posting explicitly said that the starting salary was up for negotiation, the percentage of women who tried to negotiate a higher salary shot up, matching the percentage of men who asked for more money. The findings suggest that most men weren't grappling with the "Is it okay to ask?" issue the way women were, so as soon as the employer took that item off the table, women could — and did — behave just as confidently as men.

Given this, it follows that employers have a responsibility to level the playing field by letting employees know when salaries are up for discussion. But obviously employers have incentives not to make such announcements. If you tell people that salaries are negotiable, more people will ask for more money, and more managers will have to wade through those conversations. But if you don't announce it, you've inadvertently fostered a gender bias, because more men will ask than women and real salaries will vary accordingly.

If you're like most people, your boss hasn't said, "You know, we should renegotiate your salary," or "I wish you'd lobby for a promotion." Since most supervisors don't invite these conversations, is there a way for women to ask for more and minimize the chance of backlash? Psychologists and negotiation experts have several pieces of advice for them.

But before I offer this advice, I have to come clean: I find the suggestions I'm about to list offensive. I wish women didn't have to pad their requests with niceties. I wish women could use the same language as men and get the same results. I wish that women didn't have to develop special strategies for asking for what equates to parity. "Capitulate and you will succeed" is a message I don't want to read or write. But I also know the realities women are facing. I'm well-versed enough in the research and the real world to know that if there's one place where women are judged most harshly, it's when they say, "This is what I want." I hope that in ten years, the advice I'm about to give will be as laughable as "Wear as few petticoats as possible if you'll be getting on a bicycle."[56] But for now, women have to be practical, and if you're a woman who feels that you need to make accommodations, it's good to know you're not alone.

So what are these pieces of advice for self-promotion? We'll start mild. First, a woman should strive to be relaxed when she goes into a negotiation. What does that mean? Women who smiled, leaned forward, and used calm hand gestures were considered more persuasive in negotiations than women who maintained a neutral facial expression, who spoke rapidly with few hesitations, and who held themselves stiffly.[57] In other words, all of the things that women do when they're feeling nervous? They're not going to help her when she negotiates with her boss. So if you're a woman who's going to ask for a raise, you need to practice. Get a friend and role-play until you can negotiate in a relaxed and friendly way. Or make a video of yourself so you can see where your nervousness is leaking through. The goal here isn't to grin blithely through anything and everything. Rudman finds that excessive smiling doesn't help women—if anything, it backfires. If your pleasantness seems fake, people wonder what you're hiding and, worse still, what you'll be like when your real emotions come through.

Now let's move on to the more contentious advice. Hannah Riley Bowles, a professor at Harvard University who studies gender and negotiation strategies, and Linda Babcock, a professor at Carnegie Mellon University and author of the popular book *Women Don't Ask*, advise women to combine two strategies to reduce the chance of backlash. If you're a woman in this situation, the first step is to emphasize that you care about your relationships with your colleagues. Remember the mother duck. You want to show that having good relationships is important to you because people expect that priority from women in general and especially from women leaders. You want to be explicit that relationships matter. Try using Bowles and Babcock's language as a starting point. "I hope it's okay to ask you about this. I have some questions with regards to my salary." That "I hope it's okay" line might sound very 1950s, but Bowles and Babcock's research was actually published in 2013. What does that language do? It acknowledges you care about the relationship you have with the person you're approaching. Then you explain the change you'd like to see, which might be a 10 percent raise, a particular dollar amount, or a paycheck at the top of the range for your position. You might also want a longer vacation or a different title. Once you've made your requests, Babcock and Bowles advise you to continue affirming the nature of your relationship. "I thought this seemed like a situation where you would have some valuable advice. My relationships with people here are very important to me. Would you be open to talking with me about this question of higher compensation?"[58]

Is this highly scripted? Yes. You'd need to modify it to meet your style, situation, and relationship with your supervisor. Does this sound a bit obsequious? Also yes. If it doesn't fit you or makes you roll your eyes, change it. Some of the early readers of this book said, "Surely not all managers want to hear this." That's true, but Bowles and Babcock found that such strategies improved negotiation outcomes for women in the lab, regardless of whether the person playing the role of supervisor was a man or a woman.[59] And that's an important caveat. These weren't real managers in negotiations; these were people playing managers in negotiations.

But women often encounter this negotiation dilemma with their real supervisors — *If I ask for more, will it cost me?* — and it makes many of them hesitant to ask at all. If that's you, now you have some language you can work with, a strategy that reduces backlash. If your supervisor says, "Don't ask for permission, just ask for the raise," I'd be tempted to say, "Thank you for that. So I don't have to faint or get out my parasol?" If you can make fun of the social norms, it can put you both at ease. But even if your boss waves away the gender expectations you're working around, and I hope he or she does, it still might help you to communicate you care about your relationship. Researchers find that many people who see themselves as fair-minded have unconscious gender biases just the same.[60]

The second strategy that Bowles and Babcock recommend for women is to legitimize the negotiation. A woman's supervisor often finds it easier to give her a promotion if there's a broader story. When a man asks for more money, he can list his accomplishments. He can point to the major account he landed or the new leadership role he's played. But studies show that women need to provide a story that goes beyond personal contributions; if they don't, the supervisor might fill in the gaps with unsavory additions, thinking, *She's only focused on herself* or *I guess she isn't a team player after all.* If you're a woman dealing with this, you can point to industry standards as a reason for a raise, or you can explain that a more senior person in the organization encouraged you to talk about your compensation. Help your supervisor find an easy way to make a case that's externally grounded. If you're frustrated, I understand. A person's exceptional performance should be all that's needed to make the case, but Babcock and Bowles found this added rationale moves supervisors past the question "Why is she asking?" (one question we can all agree shouldn't be on their minds in the first place).

Bowles and Babcock found that women who combined these two approaches — saying "I value this relationship" and offering external justifications for what they wanted — reduced backlash and had negotiation success that rivaled men's. Using one strategy without the other wasn't nearly as persuasive as the two strategies in tandem. When women have conversations like this, they have the option of using these tools. I hope knowing this language and being aware of the impact it can have gives women one more way to turn up their confidence.

Perhaps this advice was more offensive than you expected, or perhaps you're thinking that that's nothing compared to stories you've heard. It stinks that women have to jump through these hoops and learn little phrases that men don't, but if they're going to get the top jobs where they can effect real change, where they can transform the system for other women, then they must recognize the system's biases and, when necessary, play to them. I want the culture to change, and I want women to be effective.

The Group Needs the Memo

Let's return to the modern-day Harvard MBA classroom, where administrators have tried to improve the imbalance for women. Of the three women from this program that I interviewed, not one thought she personally needed that coaching on hand-raising. One student was the quiet type, the kind you might assume would benefit from such coaching. In fact, before she'd started the MBA program, she'd been told by a previous employer that she needed to speak up more often in meetings. But she said she'd learned how to speak up in class not from Harvard's participation seminars but from the cold-calling that happened in class — when a professor called on her unexpectedly and she had to think on her feet. She'd discovered that no matter what came out of her mouth, someone in the room could benefit from it. Another student admitted she'd always been bold. She'd had military training before entering the MBA program, and although she found the Harvard Business School classroom an extremely competitive environment, she was used to being told to put her hand down. Yet regardless of whether they started out the program on the confident or timid side, all the women remarked that they earnestly believed the coaching had helped *other* women. The recurring message was "It's a valuable thing, I just didn't need it."

I was intrigued by this discrepancy. It could be that plenty of women believed these hand-raising demonstrations helped them translate the confidence they felt into the confidence they needed to project or that the seminars provided a survival tip in a fiercely competitive environment. Or it could be that no one, at least no one at Harvard Business School, was going to admit that she needed someone else to show her how to raise a hand.

Imagine other ways that Harvard administrators could have tried to remedy the inequities. They could have identified individual students who weren't participating enough the first month of classes and sent them personalized invitations to receive one-on-one coaching on how to be more assertive and confident. This is what many managers might try first — call someone whose voice you'd like to hear more often into your office, encourage her, offer a few pointers, and then send her on her way, hoping that gives her the confidence she needs.

Would individualized confidence coaching have been as productive as holding a seminar? I suspect not. If some students had been invited to meet with a class-participation coach, those singled-out students would have felt extra pressure to perform. Quiet students already felt pressured because half the grade rested on speaking up. And I suspect most students with Wall Street résumés don't need to be told how to get another person's attention.

But the seminars? The seminars sent a message to the group, and as we saw in the second half of this chapter, it's the group message, for men as well as women, that's crucial if we want women to feel comfortable showing as much confidence publicly as they feel privately. When students came together in the class-participation seminars and heard "We expect all of you to raise your hands like this," some might have found it insulting. But the administrators had, perhaps unknowingly, tapped into something powerful. They were saying, in essence, "We expect all of you to self-promote. You'll be rewarded, not penalized, for doing so." They publicly set a standard of ideal behavior across gender lines, freeing up women to do what the men were already doing and promising women they wouldn't be punished for it.

When I first read about Harvard's hand-raising seminars, I was insulted. But now that I better understand the importance of the group's expectations, I would ask them to go a step further: Make the class-participation workshop mandatory again so that men as well as women get the message. If we want women to feel empowered displaying confidence publicly, we need to be explicit in the way we invite their contributions, whether that be in workshops, meetings, or company-wide e-mails. The message we send to the group is more important than the prodding or encouragement of individuals. The lesson we can take from Harvard is that women will behave confidently once they know they won't be penalized for doing so. Women need to know that the whole group got the memo—and that management signed off on it.

Well-calibrated confidence is where many women have a deciding edge. All of us, men and women alike, should see confidence as something we can turn up or down as needed. But it's a lot easier to turn your confidence up for the sake of pushing your agenda than it is to turn your confidence down when you're absolutely convinced you know better than most of the people in the room.

Once a woman makes a decision, she still needs to be Goldilocks in the self-promotion department—not too much self-promotion and not too little—and that can be frustrating, tiring, and unfair. But women should recognize that a mixture of confidence and skepticism is not something to self-help away. It's a powerful asset they bring to smart, insightful decision-making.

Of all the facts in this chapter, the one that bothers me most is that women apply for jobs only when they meet 100 percent of the listed job qualifications while men apply when they meet a mere 60 percent. Women need to take a lesson from the mothers of the world on this one. Women I know say that no matter what you do, you never feel 100 percent prepared to be a mother, and you just have to trust that you'll learn what you need to along the way. I hope women will learn to apply this approach to their professional lives too.

Chapter 4 at a Glance: Women's Confidence Advantage

THINGS TO REMEMBER

1. Men and women alike find it hard to believe that they're average.

2. But compared to women, men tend to think they're much, much better than average.

3. The confidence gap is largest on skills that society considers masculine or feminine. Women underestimate their knowledge and abilities on masculine skills, but men still believe they'll be proficient in skills deemed feminine.

4. Problem: Much of the professional world is considered masculine territory.

5. Overconfidence in oneself is a major obstacle to smart decision-making (examples: the two o'clock rule and CEOs who paid too much for companies).

6. Women's more accurate self-assessment means they're poised to make prudent decisions and avoid overconfident errors in judgment.

7. Confidence doesn't signal that you're right or even that you're on the right track. It signals only that you've told yourself a good story.

8. Women who are privately confident often feel they must be publicly modest.

9. There's frequently a backlash against women who self-promote. Rudman's research showed men who self-promoted were deemed more hirable, while women who self-promoted were less likable and, therefore, less hirable.

10. Once employers announced that they expected people to negotiate their salaries, women were just as likely as men to ask for more money.

THINGS TO DO

1. Think of confidence as a dial, something you can turn up or down.

2. You need to keep your confidence turned down when you're making a decision, when you're weighing your options and need to hear more.

3. Dial up your confidence when you've already made a decision and the task at hand is more about persuading others to accept it and follow you.

4. Your body, your self-confidence. To provide a quick injection of confidence, strike a high-power pose or lower the pitch of your voice. To lower your confidence, conduct a premortem or assume a low-power pose.

5. If you're a leader in your organization, make standards around self-promotion explicit; if you don't, men have reason to advocate for themselves while women have reason to hesitate.

5

Stress Makes Her Focused, Not Fragile

JUNE 8, 2015. A sunny, warm afternoon in Seoul, South Korea. Women scientists, engineers, and journalists gathered for lunch to discuss the impact of women in science. It was a few hours into the first day of the Ninth World Conference of Science Journalists, and although three women spoke at that women's luncheon, few people around the world took note of what they said. But many of us know the infamous comments made by the man who followed. "Let me tell you about my trouble with girls," he started. "Three things happen when they are in the lab: You fall in love with them, they fall in love with you, and when you criticize them, they cry." Sir Tim Hunt, who won a Nobel Prize in 2001 for his work in biochemistry, went on to say that perhaps the best science would be done if men and women worked in separate laboratories.[1]

Many people have mocked Hunt for saying that women are temptresses. Women scientists from Africa to the Arctic have chimed in, saying they feel anything but distractingly sexy when they lower their safety goggles over their eyes, tug on their vinyl gloves, or step into their hazmat suits.

But it's the last part of his message that concerns me.

Hunt said that when girls are criticized, when they're put in a highly stressful situation, their feelings get the best of them. Though he called them girls, he's not talking about moody thirteen-year-olds who didn't make the track team. It's safe to assume that if they're in a top-notch lab working alongside Hunt, the women he's referring to are experts and professionals, women with PhDs or MDs.

What does society expect of a stressed-out woman, even the kind with years of experience and world-class credentials? The message we hear, even from people more politically correct than Hunt, is that women fall apart when the pressure is on, that most women become messy and unpredictable bundles of emotions. We expect women to get emotional and irrational in intense situations. To be fragile and easily broken. But men? Society tells us that men hold it together, that they remain cool and rational. Put men in the toughest circumstances, we're told, and they can be counted on to remain clear-headed.

The surprising thing this chapter demonstrates is that when it comes to decision-making, women are actually calmer in stressful situations than men. Steadier. As we'll see, women don't become unpredictable messes. Quite the opposite. When the pressure is on, women become much more impressive decision-makers than they've been given credit for. What's most interesting is that men and women have opposite reactions when they face choices in tense circumstances. It's not that he's rational and she's not; it's that he's risk-hungry and she's not. If we want good judgment and insightful choices in stressful circumstances, the worst thing we could do is separate men and women. In order to balance each other out, both sexes have to be in the room when the pressure is on and hard decisions need to be made.

Too Emotional to Lead Versus Just the Man We Need?

As one might expect, especially at a conference crawling with journalists, people on social media erupted around Hunt's comments. Many pointed out that what he said revealed only that he's a curmudgeon. Others said he was a dinosaur and that thankfully his kind of sexist thinking was almost extinct.

Were Hunt's words a rare sighting of a nearly extinguished belief that was emboldened to dash out and show its true colors one last time? My research suggests not. It's true that Western society has been saying that women are emotionally ill-equipped to handle tough situations for centuries. The word *hysterical* was first used in the 1600s to describe women who showed signs of emotional disturbances and feeble "moral and intellectual faculties." Doctors, all men at the time, thought such intense moods were brought on by a malfunction of the uterus, which is why the word *hysteria* comes from the Latin word *hystericus,* "of the womb." Although doctors could have diagnosed comparably emotional men as hysterical, they rarely did.

Today, the word *hysterical* is less gendered and is often used to describe someone who is funny rather than irrational, but we still have plenty of other ways to say that a woman's emotions are a handicap. When Senator Hillary Clinton was running for president in 2008 and her voice broke during a speech in New Hampshire, news analysts said she was breaking down, noted that her emotions had gotten the best of her, that "she lost her cool," and they questioned whether she was "too emotional, too sensitive or too weak."[2] Professional women athletes, such as six-time Wimbledon champion Serena Williams, are regularly asked if their emotions got in their way during a competition.[3]

But is the same language directed at men who get emotional on camera? Rarely. More often, the media assures the public that tears are a testament to a man's character. When President George W. Bush cried at his 2002 inauguration, *Newsweek* reported, "He has learned not to brush the tears away," and in 2007, when photographs showed his eyes welling with tears as he presented a Medal of Honor to a slain war hero's family, the *Washington Post* reported, "The pictures were just what the White House wanted." And when male NCAA basketball players cry during March Madness, it's called "painful but heartwarming."[4] Few reporters lift the microphone to a seven-foot male after the game and inquire, "So, I have to ask, for fans everywhere, were your emotions the problem?"

If it was simply that the media disapproved of women's strong emotions but rationalized men's, that would be one problem. But there's an even more vexing issue. When women express strong negative emotions in stressful situations, they are portrayed by the media as incapable of good judgment. A woman's sadness or anger is seen as a reason why she can't lead. When Hillary Clinton's voice cracked on the campaign trail in 2008, the press questioned her ability to handle a national security crisis. News contributor Dick Morris said, "I believe there could well come a time when there is such a serious threat to the United States that she breaks down like that," adding, "I don't think she ought to be president." *Newsweek* reached back twenty years and compared Clinton's moment to Congresswoman Pat Schroeder's display of emotion in 1987. Journalist Karen Breslau wrote, "There will no doubt be comparisons to the teary press conference former Colorado representative Pat Schroeder held to announce that she wouldn't run for president, thus confirming that anyone who needed to carry Kleenex in her purse was unfit for the highest office in the land."[5]

"She's being emotional" has become the go-to dismissal when someone is trying to discredit a strongly worded decision made by a woman. Consider the military interrogation techniques of the United States, a topic that has received scrutiny from all angles. Dianne Feinstein is a female U.S. senator from California, and she headed a Senate Intelligence Committee reviewing the interrogation techniques used from 2001 to 2013. After several weeks of meetings, in the spring of 2014, Feinstein stood in front of news cameras and explained that the committee voted to declassify a report "to ensure that an un-American brutal program of detention and interrogation will never again be considered or permitted." When she announced that decision, the former CIA and NSA director Michael Hayden could have attacked many things, but he went for her emotional stability. Hayden quoted the line above and said it revealed Feinstein's "deep emotional feeling" and how those feelings were preventing her from being objective.[6] He didn't call her hysterical, but he came close. What's remarkable is that Feinstein wasn't even voicing her own personal opinion — she was reporting on the committee's deliberations. And yet Hayden attacked *her* personally.

A very different picture emerged when a male U.S. senator critiqued these same interrogation techniques. John McCain is a senator from Arizona, and he has used much stronger language, the blood rising up from his neck as he denounces certain U.S. interrogation techniques. Whereas Dianne Feinstein simply said the program was "brutal," John McCain has gone further, saying some methods were "indisputably torture" and that waterboarding amounted to "mock execution." But few criticized McCain for feeling too much.[7] Some of McCain's opponents said that he didn't understand modern interrogation techniques, while others said he was overlooking the information obtained from those interrogations.[8] They could have called him overly emotional, but they didn't. They challenged his argument, not his character.

Yvonne Abraham, a columnist for the *Boston Globe,* wondered what the reporting would look like if male politicians were treated the same way women so often are. Political candidate Charlie Baker had begun to cry during a debate for the 2014 Massachusetts governor's race, and many of the local news stories that followed expressed how it humanized him, that "Charlie Baker won the debate by losing it — by losing his composure, that is."[9] So Yvonne Abraham applied the same harsh judgment to Baker and to men's ability to lead that the media so often applied to women: "It's a dangerous world. When the red phone rings at 3 a.m., do we really want a fragile soul like Baker answering it?"[10]

Framed this way, the answer is no. No one wants a fragile leader. But that's how the media portrays expressive women in stressful situations.

I should clarify — it's not that tears are professionally damaging only to women. If a man is found crying in his office because he's heard rumors about layoffs or because his computer crashed and he lost a week's worth of work, he's judged too. In fact, a team led by Agneta Fischer at the University of Amsterdam found that a man incurs harsher judgment for crying on the job than a woman does.[11] Both are labeled highly emotional and incompetent, the man slightly more so. The problem is that women's emotions are expected to flare up and get the best of them. Debilitating emotions aren't acceptable for men, but they also aren't expected. A 2015 study found that in the United States, Republican voters often thought that men were "better suited emotionally" for politics and leadership while women were seen as "emotional," which is one possible reason that Republicans vote more male candidates into office.[12] Unfortunately, this view of women isn't limited to a political party. See a man cry and *he's* emotional. See a woman cry and *women* are emotional.

She's Emotional but He's Just Having a Bad Day[13]

So let's look at what happens in your own everyday interactions. Most of us don't see our colleagues crying at their desks on a daily basis, but we do witness milder negative emotions. We watch a supervisor frown when she learns a project is behind schedule and we hear a team leader vent behind closed doors when he's criticized. When you see a woman striding down the hall after a stressful meeting, there isn't any media coverage, no biasing headlines. What runs through your mind when you're left to come up with your own interpretation?

Neuroscientists are finding we take a much harsher view of stressed-out women than stressed-out men. Lisa Feldman Barrett, at Northeastern University, and Eliza Bliss-Moreau, at University of California, Davis, conducted a clever experiment where they showed subjects the faces of men and women expressing negative emotions. Some of the faces were sad, some angry, and others fearful, but they all looked stressed and under duress. The scientists used a program called MorphMan to ensure all the sad photos looked equally droopy, all the frightened photos looked equally alarmed, and all the angry photos looked equally frustrated. If the image that pops into your mind is a black-and-white photo of a woman with her teeth bared with rage, then you're probably remembering the classic photo from your college psychology textbook. Cut the intensity by half. These were much milder versions; instead of bared teeth, the angry individuals had tight lips and squeezed eyebrows. If you saw this expression on a stranger in an elevator, you'd recognize it as anger and you might give that person a little extra room, but you wouldn't decide to get off the elevator at the next floor.

Then Barrett and Bliss-Moreau added a brief sentence below each photo explaining why the person was stressed. According to the angry face's caption, the person "just got yelled at by the boss" or "did not get promoted at work." The sad faces had similar captions, explaining this person "just got some bad news" or that one "was disappointed by a lover." (Although Barrett and Bliss-Moreau did their research years before Tim Hunt made his infamous comments, humorously enough, they covered two of the very situations Hunt mentioned, the jilted lover and the critical supervisor.) The researchers asked the people participating in the study to imagine each person in the stressful situation so they could remember it later.

After the subjects had seen all the faces once, it was time for the second half of the experiment, where judgments were made. Barrett and Bliss-Moreau presented all of the stressed-out faces a second time, but without any captions. This time through, the participants were asked to decide, as quickly as possible, whether the person in each photo was "emotional" or just "having a bad day." Why did participants have to decide quickly? Barrett and Bliss-Moreau wanted to test their snap judgments, the kind of judgments most of us make all the time. If a waitress gives you an impatient look, you mumble to your breakfast companion, "She's a handful," or "I wonder if they're short-staffed." If the guy in front of you stares out the bus window looking dejected, you think, *He seems awfully mopey* or *I wonder what just happened.* When you see that angry person on the elevator, a harsh or forgiving interpretation runs through your mind.

So how did everyone see the stressed-out faces? The participants had just been told each person had a reason to be upset, so we'd expect identical judgments across the board: all bad days. She's sad because she just received bad news and he's angry because he was yelled at — they're both in a tough spot. And for the photos of the stressed-out men, people did chalk up their expressions to the frustrating situation. They thought that most of the men, whether sad, angry, or fearful, were probably just having bad days.[14]

But for the pictures of the stressed-out women, participants were more likely to downplay the circumstances and overplay her personality. The common snap judgment for the women wasn't that this was a passing, temporary reaction to something that had just happened. Whereas the men were having bad days, women with identical expressions were labeled emotional. Most people perceived the sad women, in particular, as emotional. At this point in the book, you're not surprised to hear that this wasn't just how the men judged the women in the photos; it's also how the women judged the women.

This sends a concerning message about how people perceive women who are in stressful circumstances. For all of us, stress shows up on our faces. When someone is picking apart a presentation you just spent all week preparing, you don't smile. At least, not until you regain your composure. When your performance review reads *Doesn't meet expectations,* you don't look or feel remotely happy. Defensive and discouraged is more like it. But even when women and men experience the same hard emotions in stressful situations, people judge women more harshly if women let their feelings show. People see a momentary flash of anger or a look of hurt feelings as a window into the difficult or emotional person she really is. As a character flaw. According to Hunt, a woman like that shouldn't be working in a lab alongside men. But an equally frustrated or dejected man? That's temporary. You've just caught him on a bad day.

It's unfair that people take a snapshot of a woman's frustration or unhappiness and judge it as a glimpse of her true nature, but when we consider what this gender bias means for decision-making and leadership, it becomes even more troubling. Whom do you want to make an important decision, the person who seems to be highly emotional or the one who probably just had a rough meeting? It's a no-brainer: You want the leader whose negative emotions are fleeting. You want leaders who keep their emotions in check and whose good judgment isn't in question.

We also learned in the introduction that people want to demote women who express anger but they want to promote angry men. Brescoll and Uhlmann's research revealed that when a job candidate is asked to talk about a past mistake, a man who shows a face of frustration is seen as deserving more responsibility, whereas a woman who shows that same face is seen as "out of control." There's a troubling pattern here: When people rebuke angry men, they challenge the content of his evaluation, but when people rebuke angry women, they challenge her ability to make an evaluation at all. It sounds as though when the pressure is on and tempers are flaring, we trust men to take the right risks. We trust that a man's emotions are coming from a sound place and that even if those emotions are strong, they won't cloud his judgment. But we believe that a woman's emotions will.

So what does the science have to say? *Are* men better decision-makers under stress? Are men making the wiser choices with a steady, unwavering hand, and are women making irrational choices we shouldn't trust?

Stress Changes Everything

Neuroscientists have uncovered evidence suggesting that, when the pressure is on, women bring unique strengths to decision-making. Mara Mather at the University of Southern California, Nichole Lighthall at Duke University, and Marissa Gorlick at the University of Texas at Austin were curious to see if and in what ways stress would change how people made decisions. They asked subjects to play a computer game in which the goal was to make as much money as they could by inflating virtual balloons. Press a button, the animated balloon got a little bigger, and you won a little money. Press the key again and you won a little more. At every step, you chose whether to keep inflating or stop and take the money. You could cash out at any time. You could go just one round, call it quits, and be guaranteed you'd win something. Of course, you wouldn't win much in a single round with a teensy balloon, and Mather and her colleagues offered a strong incentive to keep going — the tenth pump was worth more than the first pump, and the fiftieth pump was worth even more.

You're probably wondering why anyone would ever stop. Here was the rub: If a balloon exploded, if you went one pump too far, then you received no cash for that popped balloon. What made the task a little maddening was you couldn't predict how many pumps it would take — it was entirely random. Some balloons popped on the third pump, and some balloons didn't pop until the hundred and twenty-third.

In a way, each new balloon was like playing an extremely low-budget (and not terribly exciting) version of *Who Wants to Be a Millionaire?* You had to decide how much risk you wanted to take. There was always the potential for more money if you kept playing, but the chances of losing it all went up as well. Of course, in the balloon-pumping game, there were no lifelines, no dramatic stage lights, and no tricky trivia questions, but you still had to keep asking yourself, *Should I quit while I'm ahead, or is a little more money worth the risk?*

Did men and women behave all that differently in the balloon-pumping game? Not when they were relaxed. Men and women made more or less the same decisions and took comparable risks when everything was calm. Men pumped a few more times than women before they called it quits (men averaged forty-two pumps while women averaged thirty-nine), but these differences weren't huge. When our heartbeats are steady and when we feel relatively calm, men and women take the same balanced approach to risk.

But add stress to the equation, and we see something different. In people's normal daily lives, stress might be caused by any number of things: an accident, an argument, or an accusation. A due date or a blind date. In the lab, Mather and her colleagues made the stressor simple: they asked half the people to immerse a hand in a pitcher of ice water. Painfully cold water: 32 to 37 degrees Fahrenheit (0 to 3 degrees Celsius). I've tried it, and the experience changes from mildly uncomfortable to distinctly painful after about a minute. People in this study kept a hand submerged for a full three minutes. It wasn't long enough to be dangerous, but it was long enough to be stressful. When a person's body tries to adapt to this discomfort, heart rate and blood pressure go up.[15] Some people bite their lips. Others jiggle a leg or squirm in their seats. Just about everyone frowns, squints, and wishes the time would go faster.

This is called the cold-pressor task, and it's used in research labs around the world as a surefire way to make people feel stressed. So half the people in Mather's study were cold and stressed. The lucky people in a relaxed, nonstressful condition simply soaked their hands in a basin of room-temperature water (72 to 77 degrees Fahrenheit, or 22 to 25 degrees Celsius). It was as though they were sitting in a spa waiting for a manicure. After both groups finished their three-minute soaks, they had fifteen minutes to dry off their arms and relax before the balloon-blowing game began.

Fifteen minutes may seem like plenty of time for your mind and body to recover, but the researchers knew better. One of the body's main stress hormones, cortisol, doesn't peak in the body until about twenty to forty minutes after a stressful experience begins. Even after the source of the stress has gone away, fifteen minutes later, your body is still reacting. If you've ever rushed frantically to meet friends for dinner at a restaurant and then found it hard to relax even after you were sitting at the table, menu in hand, then you know that stress doesn't vanish just because the challenge has passed. Stress lingers. Even if our minds have moved on, our bodies typically haven't.

Sure enough, when the balloon-blowing game started, the stressed men and women played it very differently—differently from each other, and differently from the lucky men and women in the relaxed condition. Whereas women in the relaxed condition had pumped their balloons all the way to thirty-nine pumps, women who had just gone through the stressful ice-water experience stopped sooner, pumping 18 percent less than the relaxed women, and cashed out their winnings. The women who had been stressed chose to take the sure win over the higher risk. Stressed men did just the opposite. They kept pumping — in one study averaging about 50 percent more pumps before calling it quits. With each extra pump, they increased their bank account, but they also took more gambles, more chances that the balloon would pop and they'd lose it all. Men whose bodies were still feeling stressed were charging ahead, taking more risks than they ordinarily would, while women who were still feeling stressed were holding back, deciding that a little extra money wasn't worth losing what they had.

What might this look like in the restaurant scenario, where you've rushed in late and stressed? A stressed woman will open the menu, pick something she knows she likes, rather than one of the unusual specials the waiter describes, and then shut the menu. A stressed man will choose one of those specials, something he's never tried, and then order an expensive, unfamiliar bottle of wine, saying, "What the heck — let's hope you get what you pay for."

Risk-Seeking Men and Risk-Alert Women

What exactly did Mather and her colleagues discover? Was it a one-time fluke, these men who took more risks under stress and these women who took the sure and steady path under the same tense conditions? Before we delve any further into understanding how society as a whole celebrates how men react under stress and criticizes how women respond, it helps to see that even the scientists use biased language when they talk about men and women. When many neuroscientists and economists discuss attitudes toward risk, they often say that men are risk-seeking while women are risk-averse. According to the *Oxford English Dictionary, averse* comes from the Latin for "turned away; turned in the backward or reverse direction." Tagging women as risk-averse suggests they are going the wrong way and that there is, across the board, a right way. A more fitting term, whether we're describing men or women, is *risk-alert*. When people are alert to risks, their ears are pricked up to hear the potential warning bells; they are watchful and vigilant. I can appreciate where the term *risk-averse* comes from—loss aversion and risk aversion are two long-standing principles in economics. But if we're trying to have an unbiased conversation about risk, we need a term that's at least neutral or, preferably, as positive as *risk-seeking*. We all want to seek things and we all want to be alert, but no one wants to be averse.

What's interesting is that men don't become just a little more risk-seeking under stress. They can become almost risk-hungry under the right conditions. Some research suggests that when the stress is great enough and the reward dangling out there is sufficiently enticing, men throw caution and common sense to the wind. Ruud van den Bos and his team of researchers at Radboud University, in the Netherlands, have found that men often become laser-focused on big rewards when their bodies tense and their cortisol levels peak.[16] Van den Bos found that men under stress gambled at much higher stakes than men who weren't stressed. These weren't habitual gamblers, drawn back to the poker table night after night. These were just ordinary men and women brought into the lab and taught a new card game. As people played, they discovered they had a choice. They discovered they could play a safe version of the game, in which they would both win some and lose some but, ultimately, win more than they'd lose. There weren't many dramatic wins or losses in the safe version of the game, but people could watch their winnings slowly go up. Or they could play a risky version of the game in which they would win a big jackpot every once in a while but they'd lose a whole lot more in the process. In the risky version, the game was rigged so that players ultimately came out with nothing or, worse yet, wound up owing money. And this big loss didn't come as a surprise at the very end. The players learned, by the middle of the game, that the risky version was, indeed, quite risky.

Which game would you prefer to play? Would you go for the highly risky version—where you get a big, gratifying win every once in a while but also watch your account drop to zero several times? Or would you go for the safe version and be sure to come out with something? When they were relaxed, men and women generally chose the safe version of the game, revisiting the risky version only for a brief thrill. When men were stressed, though, they couldn't stay away from the risky decks. Van den Bos found the most stressed men drew 21 percent more cards from the riskiest deck and that the growing tally of losses didn't deter stressed men the way it deterred stressed women or calm men. The highly stressed men lost more money than they won.[17] Why would the stressed-out men keep diving in, going for the risky decks, when they kept losing? Because every once in a while there would be the reward of a big win, just enough to lure them back for more risk-taking. When the pressure is on and there's the glimmer of a highly rewarding outcome, men take gambles, both bigger and more frequent gambles than they ordinarily would.

There's also a question of who adapts their strategies as the stress mounts. In one study by Stephanie Preston, a neuroscientist at the University of Michigan, she and her colleagues told people that in twenty minutes, they'd have to give a talk and would be graded on their speaking abilities. Anyone who has ever had a boss say out of the blue, "Could you be ready to present at the meeting in twenty minutes?" knows that this is a heart-pounding predicament. But first, the people participating in this experiment had to play a gambling game, the same card game that van den Bos used, and try to win as much money as possible. Stirred, anxious, and distracted by the upcoming speech, both men and women initially had a hard time figuring out which decks to draw from.[18]

But the closer the women came to the stressful event, the speech they had to give, the better their decision-making became in the card game. Objectively better. Women tended to become more strategic under stress, looking for smaller, sure successes rather than risky prospects, racking up more money. Not so for the stressed men. The closer the timer ticked to zero, the more imminent the speech became, the more questionable the men's decision-making became; they risked a lot for the slim chance of a big win. Mind you, a big win wouldn't get them out of the speech. Taking risks was its own reward.

The pattern is a telling one: Men tend to become risk-seeking when their stress goes up, and the most stressed men can become so risk-hungry they decide the mounting costs are worth it. Just the opposite happens for women. When tensions rise, women appear to have their fill of risk more quickly.

Cortisol Pushes Stressed Men and Women in Opposite Directions

Why are men drawn to higher-risk choices under stress whereas women are drawn to certainty? Cortisol seems to be key. As you'll recall, Mather and her team waited fifteen minutes before starting the balloon-popping game so that cortisol levels had time to peak. Most of us know cortisol as the "stress hormone," as it's often described in the popular press, and that's a fair nickname. Cortisol is a steroid that the body produces as a way to cope with stress. To understand the role that cortisol plays in decision-making, we first need to understand what happens when the body sends the stress alarm.

When you feel threatened, when the police call and tell you they've arrested your son or when a colleague asks you to lead a meeting in front of the executive team with only ten minutes' notice, your body takes action immediately. Your body prepares for the well-known fight-or-flight response, in case you need to mount a defense or exit the boardroom. The first wave of the fight-or-flight response is a rush of adrenaline, which is released by individual nerve cells throughout your entire body. Fueled by adrenaline, your heart beats faster, you take in more oxygen, and you feel your muscles twitch as your arteries open up. Your face gets red. You're ready to move. You're not quite as ready to sit, which makes it considerably harder to drive to the police station or make a presentation at a business meeting.

Whereas adrenaline is on the scene immediately, cortisol is part of the second wave of your body's reaction to stress. Cortisol is slowly released into the bloodstream by the adrenal glands, peak-shaped organs that sit on top of your kidneys like two ice cream cones that have been dropped on the sidewalk. This second-wave hormone helps your body cope with stress by ensuring that you'll have the energy you need to react. If you're still responding to stress twenty or thirty minutes after the initial threat, if you're still driving to the police station or still making that presentation, you'll need more than just a faster heart rate. You'll also need stamina. But the adrenaline that your body pumped out in the first wave shut down your stomach and small intestines, so there's no fuel coming from your digestive system. Cortisol makes sure you have the blood sugar you need to think and function by prompting the liver to provide glucose in ample supply.

Men and women both produce cortisol in response to stress, but Mather, van den Bos, and several other scientists have found that the same chemical leads to opposite behaviors in men and women. If you're a woman and your body floods with cortisol, you become more risk-alert. But if you're a man and your body surges with cortisol, you become more risk-seeking. One chemical, two very different reactions.

What does this mean in practical terms? Imagine that you receive a note from your boss that reads *Come see me now*. No explanation, no context. This is not the way your boss normally operates, and you feel a rush of worry. Are you in trouble? Is someone on your team in trouble? Under these circumstances, a man usually becomes more risk-seeking, but we don't expect him to send a company-wide e-mail denouncing the boss's incompetence. Likewise, a woman usually becomes more risk-alert, but we don't expect her to go hide under her desk. It's not that anyone becomes oblivious to social norms.

More realistically, high levels of cortisol in your system would alter the choices you make once you start talking with your boss. You reach the office; he's on the phone, so you have to wait in the doorway. The minutes crawl by. Finally, he motions you in and explains, "I'm sorry to call you in like this, but my father's had a heart attack and I booked a ticket to fly home. What can you take over for me?"

The immediate threat has passed. You feel relief. But your body still has cortisol surging through its system and it will for a while, and that physiological response shapes which choices appeal to you. The research predicts that women will be drawn to more certain wins. They'll offer to take on those tasks they know they can do well, projects that are solidly within their skill sets. By contrast, the research predicts that men will be more drawn to the projects with the biggest potential payoffs, that they'll have much less concern for whether or not they might fail. They'll offer to take on tasks that are more of a stretch, especially if those projects are high-profile and could open doors. This could help explain why, when men and women are asked what they value in a job, men are more likely than women to prioritize "the potential for advancement."[19]

You might be thinking, *But I know people who don't show this pattern. I know men and women who seem unperturbed by even the most stressful situations.* Chances are you do. According to Ruud van den Bos, high levels of cortisol make all the difference. Van den Bos sees people falling into two categories: high responders, whose cortisol levels increase by 50 to 250 percent under stress, and low responders, whose cortisol levels might increase as little as 20 percent when they encounter a stressful situation.[20] The low responders' bodies are still sending out a stress alarm, but it's like the beeping of the microwave when your food is done — it's there, but it's not that urgent. The high responders' bodies, however, have a full-scale, evacuate-the-building alarm. Van den Bos and his team found that it was the high-responding men who became risk-seeking and the high-responding women who became risk-alert. Low responders remain more consistent in their approach to risk, making the same kinds of decisions in both relaxed conditions and pressured ones.

How many of us are high responders, people whose bodies ramp up into emergency mode and whose choices swerve sharply under stress? It's hard to know. Van den Bos drew a line down the middle and classified half the people he tested as high responders and half as low responders.[21] So when you look at a crowd of people, you'd expect half the women to be markedly risk-alert under stress and half the men to be markedly risk-seeking, while the rest of the men and women won't show much change in their decision-making strategies. But it may depend on which crowd you're in. Does a particular kind of environment appeal to high responders? It could be that people who show big swings in cortisol find that surge exciting and are drawn to high-pressure careers, such as news reporting or air traffic control. Or it could be that high responders are so overwhelmed by their cortisol levels that they are drawn to more predictable, calmer fields, such as optometry or massage therapy. This research on high and low responders is too new for us to know for sure.

This Is Your Brain on Stress

So high levels of cortisol are one part of the story, but there has to be more. There has to be a reason that men and women behave so differently, some explanation for why they make such different decisions under the same stressful circumstances. The same steroid is surging through their systems, yet they are drawn to different rewards. He finds the risks fulfilling while she feels satisfied by the cautious approach. Could the explanation lie in the brain? Could one set of brain areas be activated when men are making risky decisions under stress while a different set of brain areas is activated when women make those same decisions?

Mara Mather and her team began to wonder. They took their balloon-popping, cold-pressor game to a new location, a lab where they could monitor the brain activity of participants while they played the game. They used functional magnetic resonance imaging (fMRI) to examine which parts of the brain were most active during decision-making. Overall, men and women showed the same brain activity during the balloon-popping task. That wasn't surprising—after all, everyone was doing the same routine. Everyone had to watch the balloon and press a button, so the brain areas responsible for vision and motor control were active for everyone.

But there were revealing differences. For participants in the stressful condition, Mather and her team found two brain areas that responded in the exact opposite way depending on whether the brain belonged to a man or a woman: the putamen and the anterior insula.

These are small brain regions that we'll probably hear more about in the popular press as scientists discover other roles they play. The putamen is named after the Greek word for the outer layer of a peach pit because it looks like one—it has the same rounded but slightly flattened shape of a pit. And just as a peach pit is in the center of the fruit, the putamen is nestled deep in the center of the brain. The putamen plays many roles, but one of its primary jobs is to initiate action and movement. It takes input that's available from the five senses as well as the body's own internal states, gauges whether it's a good time to act, and, if it is, tells the rest of the brain, "Act and act now." When you're merging onto a freeway and have to decide whether to gun it or wait for an opening, it's the putamen that says, "Floor it."

The anterior insula, by contrast, is part of the cerebral cortex, the intricately folded pale gray layer that covers the rest of the brain like crumpled tissue paper around a delicate package. The anterior insula, like the rest of the cerebral cortex, is believed to have evolved more recently than the putamen, and it allows humans to think and act in more complex ways. When the older parts of the brain are quick to say, "Floor it," the newer parts of the brain are judging whether that's really the best call. One of the roles of the anterior insula is to announce emotions during a decision. When a person makes a risky choice, the anterior insula sends out a signal, loud and clear: "Damn, this is risky." When you floor it from the ramp onto the freeway, depending on how quickly your car accelerates, how rapidly traffic is moving, and how successfully you merged on your last attempt, your anterior insula may be squawking loudly, telling your body to be on high alert because what you're doing may get you in trouble, or it may be quiet, assuring your body that what you're doing isn't a risky choice after all.

Mather and her colleagues found that when stressed-out men needed to make a risky decision, the putamen and the anterior insula both went on high alert. These two brain regions became extremely active, which means that men were thinking, at some level, both *Act and act now* and *Damn, this is risky!* What does it mean that the emotion announcer, the anterior insula, was highly active for men? It could mean that men were having highly emotional reactions to the risky decision, contrary to the way men are usually portrayed. In contrast, women showed the opposite response. When women faced the same risky decision after a stressful experience, these two brain areas markedly quieted down. It's as though women were, without conscious intent, thinking, *No need to rush this,* and *Let's not take risks we don't need to.* Compared to men, they weren't feeling the same agitated, internal pressure to make hasty, risky choices.

Now we're getting somewhere. Men and women are asked to make the same decision under the same circumstances, but their neural machinery treats the decision and the risks very differently. His brain sends a signal of urgency while her brain sends a signal to take her time. His brain says, *Let's act, let's face the risk,* while her brain says, *Let's try caution first.*

But what's evident from this research is that men are not as full of equanimity as they're made out to be. With cortisol shaping how all of us decide under stress, neither men nor women are as clear-headed as they might like.

Beyond the Boardroom and into the Aquarium and the Rat Cage

All of this would be interesting enough if it applied only to humans. But there's one more twist to the story. We're not the only animals to have this reaction.

Andrew King, a zoologist at Swansea University in the United Kingdom, led a team of researchers that looked at how stickleback fish respond to decision-making under stress.[22] Stickleback fish are about three inches long and are named after the sharp little spines sticking out in front of their dorsal fins. With pointy spines protecting them, what do these fish have to worry about? Usually nothing, but dip a big net into their normally calm environment and you've created a stressful situation, the fish-world equivalent of a corporate buyout.

King tested how sticklebacks in an aquarium responded when he lowered a net into the center of their home. Who would swim out to investigate the net? The net was an unknown. From the stickleback's perspective, something new had just plunged into the fish tank, and it could be dangerous or it could be dinner. It could be a mate. So who was bold enough to take the risk because the goodies, those unknown goodies, might well be worth it? Mostly males. The majority of the female sticklebacks decided to stay on the periphery, in the protection of the tall plastic grass. Watching and waiting, minimizing the risk.

Researchers have observed similar patterns in other animals. Ruud van den Bos and colleagues Jolle Jolles at the University of Cambridge and Neeltje Boogert at the University of St. Andrews have been studying how male and female Wistar rats make decisions in different situations. If you see a white lab rat in a movie, its pink nose sniffing the air as it hustles through a maze, it's probably a Wistar. Compared to other laboratory rats, Wistar rats are highly active. They explore. They like to move. Van den Bos and his team wanted to see whether male and female rats would react differently to stress, so they built a large cage with a small hut on one side, a hut just big enough for a Wistar to hide in.[23]

On day one, the researchers carefully set the rats in the testing cage. The Wistars, being active rats, sniffed around and explored every corner of it, including the hiding hut, and got used to the new environment. Then the researchers put the rats back in their home cage for the night. On day two, van den Bos made the testing cage stressful by hanging a small piece of towel in one corner of it. Towels aren't normally terrifying for rats, but this was no ordinary towel. A cat had slept on this towel for three consecutive weeks.

What happened on day two when the rats were placed back in the testing cage? The female Wistars crammed themselves in the little hut, tucked in their tails, and stayed still. With an attack seemingly imminent, they spent most of their time completely hidden away. Now, the male rats didn't exactly stride over to the towel. Unlike the net that the stickleback fish were curious about, the towel held no ambiguity for the rats — that was a predator they smelled. The male rats did take cover, but only partially. These rats kept their heads out and could be seen from above. Easily visible, male rats took a bigger risk than female rats under the same stressful conditions.

Small fish and lab rats may not make many complex decisions, but when they do, they show a similar pattern to humans — males and females approach risk differently under stressful conditions. Detect a threat and she finds ways to minimize her risks, but he does not, at least not to the same extent.

Why are males and females reacting so differently? Why does she opt for the safest route as he keeps his nose out, sniffing for risk? We've been taught that when animals and people feel threatened, they have a fight-or-flight response. If they decide to fight, they lift their fists or unsheathe their claws, and if they opt for flight, they dash for the nearest exit. One could say that when the fish darted out to the net and when the rats stuck out their heads, they were opting to fight. Likewise, one could say the females chose flight when the female sticklebacks stayed in the reeds and the female Wistars crammed every last whisker into the shelter. Males fight, females flee. End of discussion.

But is it that simple? Not really. It might explain what the fish were doing, but even the male Wistars ran for cover. Flight makes a lot of sense when you're a four-inch rat facing what smells like a full-size cat. And the notion that males prepare to fight while females prepare for flight doesn't describe the dynamics among humans either. Take the balloon-popping task. When women were stressed by the icy cold water, they weren't removing their arms from the freezing temperatures in the first minute. They went the full three minutes, just like the men. And when the balloon-popping game started, women didn't stop on the first round. They didn't bolt. They could have — the researchers told all the subjects they could stop at any time. Instead of flight, the women chose to keep playing but adopted different strategies than the men, going for wins that were smaller and guaranteed, maximizing their wins while minimizing their risks. Fight or flight doesn't fully explain how males and females differ when their hearts start pounding.

Shelley Taylor has an idea about this. Taylor is a health psychologist at UCLA. She proposes that "fight or flight" describes the ways that males of many species, humans included, respond to highly stressful situations. Sometimes males choose to fight and step boldly toward the risk; the men in the balloon-popping tasks go for the bigger, more costly wins, and the male fish dart to the middle of the aquarium. And sometimes males choose flight, just as the male Wistars ran for cover.

But Taylor believes that there's another way to describe how females respond to stress. Instead of fight or flight, Taylor says, females "tend and befriend."[24] She argues that from an evolutionary standpoint, it makes sense that males make more aggressive moves under stressful circumstances. Males boldly size up the threat and if they think they can win, they attack, but if the risk turns out to be too great, if darting out proves to be a terrible idea, then fleeing is there as a backup. If males can always choose flight, the risk is worth eyeing and often worth taking.

But females don't always have flight as an option. Females of most species are charged with taking care of the young, with tending them and bringing them to maturity. They invest more in their offspring than most males do, so they can't exactly flee under stress because that would mean leaving their young behind. Likewise, fighting would mean turning their backs on their young to face their attacker, once again leaving the young unprotected. So fight-or-flight isn't going to work for females.

Under threat, females need a strategy that minimizes risk rather than embraces it, a default approach that involves staying in the threatening situation but increases the chances that they'll survive and so will their young.[25] So, Taylor argues, they find allies. If a female befriends others, if she creates a strong network, chances are that more members of her group will protect her and her young when there is a threat. It's good for that individual female because she has protection, and it's good for the species because other adult females might have young to protect. Playing it safe is okay because she won't be alone.

You might be wondering whom they befriended. In all of these studies, there wasn't anyone to befriend. The fish didn't have any girlfriends in the tank, the rats didn't have any pups, and the women didn't have anyone to cheer them on. It's interesting that there doesn't need to be someone to huddle with—evidently, it's not the presence of someone else that triggers this risk-minimizing response. The risk-minimizing strategy seems to be triggered instead by a highly stressful event. The female sticklebacks stay in the reeds when a net suddenly appears, and the female Wistars stay in the hut when they smell a predator. Minimizing risk is their best option.

The stressed women in the balloon-popping task didn't hide, of course, but they did find a more sophisticated way to minimize risk: they chose to win as much as they could, but they also chose to stop before they lost what they had gained. The stressed women backed away from risks sooner than the stressed men, but the stressed women still took gambles. They just walked the middle ground.

This is a fascinating way to think about gender differences in decision-making under stress. Males may have evolved with the deep, unconscious belief that they can always walk away (or run away, if need be). His decisions can vary more; he can afford to be risk-seeking when he encounters a threat. But females historically haven't had that luxury. Because turning and running hasn't always been an option, perhaps females have evolved to be more steady and cautious in their decision-making. *Do the safest, surest thing,* their brains coo when their hearts start pounding and cortisol surges, *because we don't have an easy escape if this goes badly.*

Some scientists are skeptical of the tend-and-befriend theory and point out that while this might be the reactive coping style for women, some women can adopt a proactive coping style. When they feel stress rising, these women step up and try to change the circumstances rather than turning to others.[26] At the same time, there is growing support for Taylor's tend-and-befriend view.[27] Most recently, a series of three studies published by University of Vienna psychologist Livia Tomova and her colleagues found that under stressful conditions, women became more attuned to other people.[28] In one experiment, people reached through a curtain and touched something pleasant, like a feather or a cotton ball, or something unpleasant, like a slimy mushroom or a plastic slug. Each person simultaneously saw a picture of what he or she was touching and a picture of what someone else one cubicle over was touching, and the subject then had to rate the pleasantness of their respective experiences. Typically, people merged the other person's experience with their own — *If I'm touching something pleasant, then I'll rate your slug-touching experience as nicer than I ordinarily would.*

Tomova and her colleagues' findings would be considered surprising to many of us. Women showed heightened attention to others and more empathy when they were stressed from having to give a public speech than when they were relaxed. Stressed women found it easier than usual to take another person's perspective. Just the opposite happened for the stressed men — they became more egocentric. They became overly focused on their own experiences when they were stressed, finding it harder to imagine what someone else might be going through. For men, the thought seemed to be *This piece of silk I'm stroking is incredible, so that cow tongue you're touching can't be all that bad.*

Stress had a tunneling effect on men's attention, and when they had to take another person's perspective, it actually slowed down their thinking considerably. This may seem insulting to men, but the way the men reacted is probably what you would anticipate from everyone. Would you really expect to notice other people's needs more when you're stressed? Only if your automatic response in a crisis is to tend and befriend.

While biology isn't the same as destiny, we do need to be aware of these propensities. We all lock onto the attractive sides of a choice when we're stressed. But what we find attractive can differ radically — whereas men find risks attractive under stress, women are drawn to the sure thing.

Having Women in the Room Matters

Is one decision-making strategy better than the other? Is it better to take big risks and focus on the chance of a huge, rewarding success, or is it better to go the safe route, where the wins will be smaller, but the risks are fewer? Ruud van den Bos uses the example of a house on fire to illustrate that neither approach is optimal all of the time.[29] Let's say someone is standing outside a burning house. The fire is contained to one small room at the back, and there's an excellent chance that he or she could safely rescue someone who is trapped inside. Because the odds are good, he says, people would probably think that a risk-seeker who runs into the burning house is heroic, while a risk-alert individual who just stands on the sidewalk is choosing poorly. But if the fire has engulfed the entire home, if the roof is crashing down and the odds of rescuing someone safely are extremely low, then as hard as it would be to stand on the sidewalk, everyone would probably agree it's the wise choice.

In the real world, we usually don't know how big the risks are. Without that certainty, we want to be sure we have both perspectives present when big decisions are being made, both the risk-seeking and the risk-alert, so that one perspective can temper the other.

What's the result when women are excluded from decision-making altogether? Let's look at what happened to companies that did and didn't have women on their executive boards during the global financial crisis. Credit Suisse examined almost twenty-four hundred global corporations over six years, from 2005 to 2011, and found that large-cap companies with at least one woman on the board outperformed the stock prices of comparable companies with all-male boards — by 26 percent. The net income growth for these companies was higher as well; during that same six-year period that was such a struggle for most firms, companies with women on their governing boards averaged 40 percent more income growth than companies with no women on their boards.[30] Much of the success is due to the fact that these companies rebounded more quickly. In the three years after most stock prices bottomed out in December 2008, companies with one or more women on their boards dug their way out of the pit faster, and their stock shares continued to be worth more even as the companies with all-male boards recovered.[31]

Some might say, "Fine, fine. The next time we have a global financial meltdown, we'd be wise to put some women at the helm. Agreed. But in the meantime, during a normal year when the stock market is rising, those women would hold us back. They'd tell us to play it safe. Under normal conditions, it's going to be the companies that take the big risks that will make the big profits."

It's an interesting point. Make sure to put a woman on the board during crises, but not during lucrative times. It's also an empirical point, one we can test by looking at the market before the crash of 2008. Did firms with women on their executive decision-making teams see lower profits prior to the crash? Not at all. From 2005 to 2007, the stock performance of companies with at least one woman on the executive board looked almost identical to companies with all-male boards. Having a woman in the room wasn't holding anyone back. Nothing lost, but much gained.[32]

Gender diversity on corporate boards is a controversial topic that has generated a flurry of debate and an avalanche of competing results. Some economists find no market differences between companies with women leaders and those without, while others argue the difference is striking. One extensive analysis seeking to reconcile these contradictions reported that where a firm was headquartered mattered. The companies that benefit most from female board members are located in countries with greater gender parity; that is, in countries where women earn almost as much as men for the same work and where women are elected into public office at fairly equal rates.[33] Having more women in the room helps only when their views are taken seriously.

The finding that women focus on sure wins under stress could help explain why female senators led the United States out of a government shutdown in October 2013. Nearly eight hundred thousand federal employees were on furlough, and each hour of the shutdown cost American taxpayers an estimated $12.5 million.[34] By day fifteen, the pressure was mounting to find a solution, and it was a group of female senators who led the way. The female senators credited their success to their having taken a more collaborative approach to decision-making than their male colleagues, to reaching across the aisle more often. But now we see women bring an unrecognized strength to the table. When the pressure is on and the situation is urgent, women aren't looking for a one-in-a-million Hail Mary pass. Women become more task-focused. They don't aim for over-the-top. They aim for achievable.

This has at least two implications for how we run organizations. The first is obvious — we need to make sure more women are in the room when big decisions are being made, particularly when the decisions involve assessing risk. As of December 2013, women made up only 16.9 percent of the executive boards in the Fortune 500, and one out of every ten companies had no women on the board at all.[35] The United Kingdom is doing considerably better; women hold 28.6 percent of the seats on executive boards in the FTSE 250 (the top 250 companies on the London Stock Exchange).[36] Of course, corporations aren't the only groups where men dominate decision-making. The website 100percentmen.tumblr.com has the tag line "Corners of the world where women have yet to tread" and lists a variety of organizations that are run exclusively by men. The week I checked, there were photos of the twenty-three leaders of the Institute for New Economic Thinking, an organization dedicated to promoting "relevant economics that serve society." All of the leaders were men.

People have been drawing attention to the low number of women in leadership for years, so the concerns are hardly new. But the reason to address them is. This isn't just a matter of fairness, or equal opportunities in the workplace, or even the broad demonstrated benefits of diversity. The reason to include women is more specific to the situation at hand. Men and women approach risky decisions very differently under stress. If more women were key decision-makers, perhaps organizations could respond more skillfully to small stresses and not allow them to escalate into huge ones.

The second implication is a bit less obvious but just as important. Once women are in the room, they need to speak up. Several bestsellers, such as *Lean In* and *The Confidence Code,* have made this point recently, emphasizing the benefits that men receive but women do not. Because men are more likely to speak up and shape the conversation, they enjoy more confidence, more influence, more respect, and more money. But there's another benefit of speaking up that I hope resonates from this chapter. Women need to speak up not just for their own personal good but for the good of the decision. If you're in a pressured decision-making meeting and no one is bringing up a risk that sets off alarm bells for you, it doesn't mean that everyone has thoughtfully weighed that risk and decided it doesn't matter. Under ordinary circumstances, it might be a conscientious group. But in a stressful situation, it could very well be that a risk that's obvious to you isn't obvious to anyone else. The only way to make sure the risks are named and evaluated is to raise them.

The Glass Cliff

Does this mean we should ask women to lead only in times of high stress? If women bring balance to risk-assessment and are more attuned to the needs of others under pressure, is the best time to ask a woman for her input when nothing else is working? No. Putting a woman in charge in a crisis is called the "glass cliff," a phenomenon first observed by University of Exeter social psychologists Michelle Ryan and Alex Haslam. In 2005, Ryan and Haslam noticed that the top one hundred companies on the London Stock Exchange were appointing more women to the highest leadership positions, particularly on their executive boards.[37] Initially, it seemed like great news. Activists had been demanding that kind of representation at the top for years, and it appeared that women were finally breaking through the glass ceiling. But there was more to the story. Businesses were inviting these highly qualified women into boardrooms only in times of crisis, after a company had weathered several months of dismal stock performance. When organizations were doing well, when their stock values were stable or rising, those companies were still choosing men, not women, to sit at the table. Ryan and Haslam named this pattern the glass cliff because women who broke through the glass ceiling were more likely than men to step into precarious, risk-laden leadership roles. And because these women were inheriting organizations in crisis, they were positioned to fail.

At first, this seemed to be a problem just for the United Kingdom. But over the next decade, the same pattern appeared worldwide. Think Janet Yellen, the first woman to head the U.S. Federal Reserve, appointed when fiscal policy was failing and the Federal Reserve had at least $1 trillion of risky mortgage-backed securities on the books,[38] and Christine Lagarde, the first woman to direct the International Monetary Fund, chosen after her male predecessor resigned under international sexual assault charges.[39] Or think Mary Barra, the first woman to become the CEO of a major automotive company. When General Motors announced that Barra would be the next CEO, many celebrated the historic moment, citing it as proof that the tide was turning for women. A woman was finally leading the ultimate boys' club, a car company. But within weeks of Barra's appointment, General Motors began recalling millions of vehicles. A few months later, documents surfaced showing that the company executives had known about fatal problems with two vehicles for years. They knew their defective ignition switch was to blame for thirteen deaths, and they had been lying to the families, to the public, and to the government.[40] The tide wasn't turning at all: people pick women to lead when things are falling apart.

One could point to this research on women becoming risk-alert under stressful circumstances and say, "But it's wise to put a woman at the helm in times of a crisis because she won't make the same mistakes a man is likely to make. She'll stay cool-headed. Almost literally. Her anterior insula and putamen are less likely to go into overdrive."

But if more women were key decision-makers in successful times, organizations could respond effectively to small stresses, not wait for the massive ones. And if women are asked to lead only when an organization is crumbling and they fail to save it, if women get to steer only once the ship is sinking, then people can point to those failures and say, "See, it doesn't work to have a woman in charge after all."

Two Options Are Better Than One

What can women do to make strong decisions even when the pressure is on? We've already looked at the benefits of ensuring that both men and women weigh in on decisions with far-reaching consequences. There's another indispensable strategy, indispensable for both men and women, but it doesn't come naturally: Resist the first idea that comes to mind. When you're under stress, take the time to generate at least two options before you decide. This will be incredibly hard to do when the time comes because when you're stressed, you usually want to move quickly from *What am I going to do?* to *At least I'm doing something.* Generating another option means hesitating. It means delaying the relief of resolution.

But when you do start moving, you're much more likely to aim in a productive direction. Why do you need to generate more options when you're feeling stressed? It's not that your first idea will necessarily be bad. It's that you'll see that first idea as better than it really is. Research shows that one thing men and women have in common when they're making a decision under stress is that they pay more attention to the rewards attached to the options in front of them.[41] As we saw with the risky decisions in the balloon-inflating and card-playing experiments, men and women focus on different rewards (she focuses on the modest but attainable rewards, he focuses on bigger but hard-to-get rewards), but they are both focusing on rewards. If you face a decision under stress and you have only one option, you're going to be more seduced than you normally would by the positive aspects of that option.

Let's say you apply for a job and they make a great offer. But they're giving you only a week to decide and that's very stressful for you. You expected to have more time to think about it and determine whether it's really better than your current job. Chances are you'll focus on the incredible perks of that new position: they pay well, it's highly prestigious, and you'll meet some of the top people in the field. Under those pressured time constraints, you'll downplay the negatives, the biggest being that you'll go from working a very manageable forty hours a week to working fifty or more.

How does your decision-making process change if you give yourself a second option? Chances are you'll still focus on the positives, but now you can compare the positives of two options, not just delight in the possibilities of one. In this case, the second option can't be "I'll just continue in my current job, same-old everything, without any change." That's disheartening. The second option has to be a proposed change as well. You could look for other job openings. Now that you know how desirable your skill set is, you'll probably see job postings differently. Or give yourself the option of asking your current boss for something that matters to you: a raise, a promotion, an extra week of vacation a year, or a day a week to work from home. You're not going to secure another job before the end of the week, and your boss may not approve a raise or whatever you've asked for that quickly, but if you explore multiple options rather than drilling down exclusively on the job offer you have, you're less likely to talk yourself into a bad choice.

Because you're improving the process, the outcome improves as well. Research shows that increasing your number of options improves the quality of the decision you ultimately make. Paul Nutt, a faculty member at the Fisher School of Business of Ohio State University, analyzed 168 decisions made by businesses, nonprofits, and government agencies.[42] He looked at choices being made by major organizations, such as McDonald's deciding whether to redesign their stores nation-wide, as well as choices being made by small organizations, such as a rural hospital considering whether to expand and add a detox unit. He found that 71 percent of the teams making these decisions considered only a single option. Most of these teams simply asked, "Should we do it or not?," just as you might ask, "Should I take this job or not?" In most cases, there was just one option on the table and all of the team's in-depth analysis revolved around whether to accept or reject that option. Nutt found that when teams asked only, "Should we do it or not?," they came up with decisions that were considered problematic 52 percent of the time over the long haul. Focusing all of their mental energy on a single option resulted in bad decisions more than half of the time. But teams that considered two or more options came up with decisions that succeeded 68 percent of the time. And that's the average. If we zoom in on individual organizations, we see how much a team benefits when it goes from reviewing one option to reviewing two. When executives at a German firm looked back on their decisions over an eighteen-month period, they found that only 6 percent of the "Should we do it or not?" decisions were rated "very good" choices in the long-term. But when they expanded their options and considered at least two alternatives, they were pleased with their choices 40 percent of the time.[43] Admittedly, these research studies looked at teams making decisions (and the German firm had high standards), but if all the members of an entire team get tunnel vision when they evaluate only a single option, an individual can't hope to fare much better.

Generating more options is going to be hard to do when you're stressed, so practice when you're calm and relaxed. Make it a habit. When you're trying to decide something simple, such as whether to accept a party invitation, in addition to asking, *Should I go or not?*, put another option on the table. Ask yourself, *Would I rather get together with this friend solo later in the month?* You have two options to compare, and another good one may follow.

Fans of Barry Schwartz's book *The Paradox of Choice* may disagree with this suggestion. Schwartz argues that too many choices stifle decision-making. When we have too many options, explains Schwartz, we run into choice overload and find ourselves not wanting to decide at all. Imagine that you wake up on a gorgeous Saturday inspired to fix the bike you haven't ridden in years. You drive over to the bicycle shop, hoping to pick up a screwdriver, and find not one but fifty screwdriver-like tools. You might understandably feel overwhelmed and be tempted to head for the door, empty-handed and rationalizing that you wouldn't know how to fix your bike anyway. That's choice overload. Schwartz finds that having too many options reduces both your initial motivation to choose and your later satisfaction with the choice you do make. But the research on choice overload is mixed. As Chip and Dan Heath explain in their book *Decisive,* choice overload usually kicks in when you have more than six options.[44] Giving yourself one or two extra options when you face a decision under stress isn't going to paralyze you.

That's Excitement You're Feeling

Another strategy that's likely to improve your decision-making in times of stress is reappraisal. We usually think of stress as a bad thing. When you feel your heart thudding in your chest right before you get up to give a presentation, you fear that you'll make mistakes or forget what you planned to say. You tell yourself, *Just calm down,* because like most of us, you think your nerves will get in your way. Alison Wood Brooks, a faculty member at Harvard Business School, is one of many scientists who finds this strategy doesn't work. In fact, telling yourself to stay calm can actually backfire.[45] Your body is clearly not calm and now you feel like you're failing on two fronts. If you can't calm down, how can you possibly give a good talk?

Brooks has a better suggestion. Rather than trying to ignore or reduce your stressful state, reinterpret your body's stress as excitement. Try telling yourself, *I'm really excited about this.*[46] I realize this sounds a bit far-fetched, given how differently we use the words *excited* and *stressed.* I'm excited to buy a new car, go on vacation, or see an old friend, but I'm stressed about meeting a deadline, taking a test, or seeing my father for the first time in four years. Excitement frames the future as a challenge you're looking forward to, while stress frames the future as a threat you might not be equipped to handle.

One's thoughts might be very different, but stressed and excited states do have some similarities.[47] Your heart pounds, your palms sweat, and your cheeks turn pink. We've all had the experience of not being able to sleep the night before an exciting event, just as we can't sleep the night before a stressful one. You could take all of your body's stress reactions and simply reframe them as excitement.

But does it actually work? Can people walk into an obviously stressful situation and just tell themselves, *What I'm experiencing right now, this is excitement,* and believe it? Brooks decided to find out. To make people stressed, she had subjects do something most of us want to do only after a few drinks: karaoke. It might be fun to perform in a bar with your friends cheering you on and plenty of background noise, but it's another thing altogether to stand, sober, before a stranger in an otherwise silent room and sing.

When paid volunteers arrived at the lab, Brooks explained they'd have to stand in front of an audience and sing the 1980s hit by Journey "Don't Stop Believin'" using the Nintendo Wii's karaoke video game.[48] She confirmed that each person was familiar with the song, then she made an unusual request. She said that when the experimenter came into the room and asked the subjects how they were feeling, they should give a specific, scripted response. She told half the participants to say, "I am anxious," and the other half to say, "I am excited." Each person was randomly assigned to a group, so it was the luck of the draw whether a person was in the anxious or excited group. Some people would be lying, of course, but Brooks told them to try to believe what they were saying.

And then the difficult moment came. They sang. They stood in front of total strangers holding a microphone and sang as best they could. Just like regular karaoke, they watched the lyrics pop up on a TV screen, but that was all the help they had.

What happened? Brooks used the Nintendo Wii's karaoke video game to rate each person's singing. The Nintendo Wii objectively and mercilessly rates your singing performance on a scale of 0 to 100 based on whether you're loud or quiet at the right moments, whether your pitch is accurate, and whether you hold a note long enough. It's not very forgiving. What's interesting is that the people who said "I am excited" were much better singers and scored much higher than the anxious people. The excited people scored 80, on average, while the anxious people scored only 52. Saying three little words — *I am anxious* or *I am excited* — was enough to change their performances, dramatically so. Keep in mind, the "excited" people didn't come into the experiment inherently more optimistic or eager. They were just asked to say it out loud and do their best to believe it. And, remarkably, they did believe. At least, they believed enough to outperform their anxious peers.

If the words *I am excited* improved karaoke singing and nothing else, knowing that might help us approach our Saturday-night outings with more dignity, but it wouldn't help most of us approach our jobs on Monday morning. But Brooks didn't stop there. She tested whether saying "I am excited" helped people overcome other common anxieties. She looked at math performance. When adults took a timed math test, the ones who were instructed to be excited solved more problems correctly than those who were instructed to stay calm. What about public speaking? Again, *I am excited* were three magic words. People who were instructed to say that to themselves before they stood up to give a speech were rated as more confident, more competent, and more persuasive by a panel of external judges.[49]

How does reinterpreting stress as excitement lead to better decisions? This field of study is rather new, so there isn't much research yet showing that reappraising stress directly bolsters decision-making. But Brooks and other researchers are finding that reappraisal improves people's thinking in several ways that help decision-making. For one thing, people are more confident in their abilities when they tell themselves, *I am excited about this.* As we learned in chapter 4, having the right amount of confidence improves decisions.

People are also less likely to latch onto potentially negative comments when they reappraise. One of the things that happens when you're stressed is you become more sensitive to negative cues. Everything looks and sounds like a threat. It's as though your antennae are out there quivering and searching for criticism. If you're stressed about your annual performance review and you get an e-mail saying, *We need to postpone our meeting until Friday,* you read that as a sign that your manager is stalling because she has to give you bad news, not as an indication that she's juggling a busy schedule. If you're feeling stressed on a first date and you hear, "That leather jacket you're wearing is very interesting," you're likely to interpret that as a criticism rather than a compliment. If you're seeing most things as threats, it's going to cloud your judgment.

But reappraisal buffers people from those threats. Two other researchers at Harvard, Jeremy Jamieson and Matthew Nock, along with Wendy Berry Mendes at the University of California, San Francisco, show that when we reappraise, we're more protected from the negative cues around us.[50] When you interpret your body's reaction to stress as something that will improve and enhance your performance, when you tell yourself, *I'm glad my heart is pounding; it means I'm ready for this,* then you see and hear the world differently. You're not as distracted by this person's furrowed brow or that person's ambiguous comment. If you're not seeing threats everywhere, then you'll make better decisions. You'll find it easier to prioritize the topics to raise in your performance review and you'll more accurately judge whether you'd like a second date.

There might be a fine line between being stressed and being excited, but whenever you can, lean toward the excited side. You'll make stronger decisions and have an easier time with challenges, and your mind will be more focused. Most of us can't stop ourselves from getting nervous, but we can stop our nerves from getting in our way.

We need to reexamine our preconceived notions about stress and gender. When men and women are ruffled by a stressful situation, we judge her as the emotional, erratic one and judge him as the stoic, steady one. The research in this chapter undermines those notions. Men under stress aren't steadier and more rational than women under stress. They're lunging for risks they would ordinarily avoid, betting more than makes sense, darting out for a quick look at what might be rewarding in that net.

What should women take from this? If you're a woman faced with a high-pressure decision and the men around you are gunning for an idea that you think sounds incredibly risky, speak up. Even if these are normally sensible peers whose judgment you've always respected, stress changes people. And as a woman under the same pressures, you're more likely to be alert to the danger, to know the net isn't always full of goodies.

When Tim Hunt joked that segregated labs might improve scientific inquiry, men and women alike were offended. We all bristled at the insult. But segregating labs isn't a terrible idea just because it's sexist; it's a terrible idea because it's shortsighted. When the pressure is on, we want women balancing out the decision-making impulses of men and vice versa. We need both sexes in the room. We can't make the high-stakes decisions in science, government, or business any less stressful. But we can ensure that when the pressure rises, there's a better balance between taking big risks and making guaranteed progress.

Chapter 5 at a Glance: Stress Makes Her Focused, Not Fragile

THINGS TO REMEMBER

1. A common misperception is that when women are stressed, they become emotional and fall apart, but men remain clearheaded and hold it together.
   - When a woman expresses negative emotions, she's seen as emotional.
   - When a man expresses those same emotions, he's seen as having a bad day.
2. Neuroscientists have uncovered evidence suggesting that, rather than swooning under the weight of their emotions, women bring unique strengths to decision-making.
3. Researchers have found that a spike in cortisol levels has opposite effects on men's and women's approaches to risky decisions.
   - The most stressed-out men pursue options that have big costs and a small chance of big benefits, while the most stressed women go for the smaller, more guaranteed successes.

4. Men's and women's brains often react differently to risky decisions under stress.
   - Mather and her colleagues found that when stressed-out men needed to make a risky decision, the putamen and the anterior insula went into overdrive.
   - But for women facing the same stress, these two areas became quieter and less active.

5. Problem: Successful women face a glass cliff and are often promoted to leadership roles only in times of crisis.
   - If more women were part of the decision-making process, perhaps organizations could respond effectively to small stresses and not allow them to escalate into huge ones.

6. Humans aren't alone in their gendered reactions to highly stressful decision-making. Wistar rats and stickleback fish also show males taking risks and females minimizing them.

7. Whereas fight-or-flight seems to capture how males of different species respond to threatening situations, tend-and-befriend seems to better describe how females react.

8. Women's responses to stress help balance out men's, so it's not just good for women if they're included in the decision-making process; it's good for the decision.

THINGS TO DO

1. Generate at least one more option when you're making a decision, even when you're stressed. You'll make better decisions with more options on the table.

2. Reframe "I am anxious" into "I am excited" and see your decisions markedly improve.

6

Watching Other People Make
Terrible Decisions

"SHE DIDN'T DO anything for seven months. Seven months." The woman sitting across from me shakes her head, trying to understand why her older sister waited so long to call a doctor. Lori is a forty-nine-year-old elementary-school teacher in Texas. She backs up and tells me how things started.

"So she first noticed a problem in June. The story she's told me is that she was hurrying to leave the house, late for a meeting she was leading at the library, but she realized she'd better use the bathroom first." Lori lowers her voice a little. "My sister hates to have a bowel movement in a public restroom — she's a little old-fashioned that way. Anyway, she goes to flush and notices what looks like a little blood in the toilet bowl. A teeny-tiny bit, my sister insists. She says she was alarmed, she swears she thought about calling her doctor, but she was already running late so she decided she'd deal with it later. She'd call when she got home. But then she became swept up in her day and completely forgot about it, until a week or two later when it happened again. This time it was right before she went to bed. Again, a bad time to call, or so she says." Lori stares into space for a few seconds. "She said there wasn't any pain at all, so she didn't think it was a big deal.

She started eating more fiber. At some point, a month or two in, I guess, it did start to hurt, but she took some Advil and that made it better. Given where it hurt and how her high-fiber diet wasn't helping, now she thought it must be related to menopause. She started taking Aleve, still not seeing her doctor. Not telling her husband. Well, at some point she told her husband, but why he didn't drive her straight to the emergency room, I don't know.

"Do you know what finally got her to go to the doctor?" Lori asks me. "She couldn't pick up her grandkids. She was in too much pain to lift them, and that finally convinced her to do something. She made an appointment, went in to see her regular doctor, and they immediately sent her to the hospital for testing. A day later, we knew — it was cancer. I got the call on my lunch break, and that's the first time I'm hearing any of this. Stage three bowel cancer, late stage three. Really bad. I looked up the prognosis, and it's only a fifty to sixty percent survival rate once it gets that far.

"I'm so angry. My brother and I both want to strangle her. But I'm also confused. This makes no sense because she's so much smarter than this. And she worked in a dentist's office for years — she knows what happens when you ignore early-warning signs."

We all shake our heads when we hear a story like this, hoping our loved ones have more sense. But most of us have watched an otherwise intelligent, sensible person make a profoundly bad decision. Perhaps a friend decides to return to her husband even though he's walked out on her several times, or a client you represent refuses to consider a generous offer. The fact that it's a terrible decision is obvious to you, but the insistent individual believes that he or she is making the best choice possible. How is it that smart people can make awful decisions but convince themselves otherwise? We're all highly skilled at rationalizing and justifying our own behavior, at saying things like "Well, I guess one bite wouldn't hurt" or "No one will notice," but there seems to be something more at play here, something that goes beyond fooling yourself a little. This chapter looks at why people are lured into fooling themselves a lot. The saying "We learn from our mistakes" suggests that all of us cherish the opportunity to review a poor judgment and learn from it. Perhaps a few of us do. But we'll see why pivoting away from a mistake requires a Herculean effort compared to making that same mistake again and again.

I should be clear — this isn't a chapter about faulty reasoning. Faulty reasoning is certainly a human problem, and we all make decision errors that go unnoticed in everyday life. As we saw in the chapter on intuition, most of us experience the anchoring effect when we're deciding how much something is worth and we latch onto whatever number we see or hear first, no matter how random or how ridiculous. We fall into the confirmation bias, discussed in the chapter on risk-taking, believing that we're "doing research" when in fact we're collecting only the information that supports whatever decision we want to make. Our reasoning may be skewed, but no one around us notices because the decisions still sound sensible.

This chapter, however, is not about sensible. This chapter is about bad decisions that are right out in the open, those glaringly questionable options people choose that make you want to cover your eyes. Everyone realizes that it's a bad choice except the person who's making it. That individual seems convinced that the decision is right, and when she discovers that others disagree, she simply stops talking about it rather than changing her mind.

And that could be the most revealing lesson of all: our reactions, our please-don't-tell-me-you-did-that recoil to other people's crazy decisions, don't always prompt change. We can believe the person gets it and learn a week later that she made the same shortsighted choice all over again. Often we misunderstand why family and friends make bad choices, and we target the wrong problem. We believe they're not thinking through the consequences of their choices, but we really need to consider their motivations — where they started, not where they are headed. The goal of this chapter is to help you understand some of the motivations that drive bad decisions. You'll be a little less frustrated if you understand what might be happening and you might have a little more faith in other people's sanity. I'll also identify strategies for targeting different motivations, for nudging people toward better choices. Let's be realistic — it's someone else's decision, so you don't have much control over what she ultimately chooses to do. But once you understand why she stepped onto a particular path, you have a chance of helping her pick a better one.

Ask "When?" Not "Who?"

We tend to assume that a certain type of person makes poor decisions, and we label people accordingly — he has terrible judgment, she doesn't have any business sense, those two are set in their ways. We treat it like eyesight and driving — some people can't see ten feet in front of them and shouldn't be trusted behind the wheel. But that's oversimplified. Thinking that some individuals shouldn't make important decisions might feel justified in the moment, but we've missed something crucial. Scientists are finding that quite often, it's the circumstances, not a person's inherently flawed way of seeing the world, that lead to bad choices. The analogy to eyesight and driving works if we extend it further: Of course some of us have better eyesight than others, but nearly everyone finds it hard to drive at night in the pouring rain. Because we know those circumstances signal bad driving conditions, we all slow down on the road. We need to learn to recognize bad decision conditions as well. We need to ask *when* people will have trouble making a decision, not who is likely to have that trouble.

As parents reach their golden years, adult children often wonder if they can trust their parents' decision-making abilities. It's common knowledge that a person's memory starts declining as he or she grows older, but most of us don't realize that the decision-making process also changes with age. But *declining* isn't the right word for it; it's more like *tilting*. As people get older, they lean toward certain pieces of information and willfully pull back from the rest.

I should also clarify that this chapter isn't about women making terrible decisions; it's about women watching others making terrible decisions and ways they can handle that better. It's not that women have a harder time with other people's bad choices. Both men and women cringe when they watch their loved ones show atrocious judgment, and both want their coworkers, friends, and family to choose well. But there are two reasons why we should think about women through the lens of how others make decisions. First, as we learned in the chapter on decisiveness, women tend to be more democratic in their decisions, asking others for their input, which means women's lives are more likely to be affected by other people's terrible judgments.

Second, women are often decision-helpers. Women find themselves (or perhaps put themselves) in decision-support roles more often than men. When a husband faces a medical decision, his wife typically becomes more anxious and prefers to play a larger role in the consultation process with the doctor than he would play if the roles were reversed.[1] When college students have questions about a big life transition, such as whether to apply to graduate school or which company to work for, they are more likely to seek advice from a female professor than a male professor.[2] Sons and daughters also play different roles. When an aging parent needs help from adult children, sons are more likely to run errands outside of the house — going to the bank, picking up the groceries, filling a prescription — while daughters are more likely to provide personal care. A daughter is probably going to be the one to help her mother or father bathe, use the bathroom, pick out clothes, pay the bills, and make dinner, which means she's going to see some of the most intimate and hidden choices her parents make.[3] She might notice that her father is paying his new financial planner exorbitant fees or that her mother has a growing number of unopened Avon products. Women who can recognize early when someone might be motivated to make a poor choice are well positioned to help persuade that person toward another path.

When Should We Be on the Lookout for Bad Judgment?

So what are those difficult decision conditions? When are those around you prone to poor judgment?

The first one is simple: People make poor choices when they have paid a lot of money for an opinion. This might happen when they've hired a consultant to help solve a problem or when they've paid out of pocket for a doctor who isn't covered by their insurance. These experts don't necessarily offer inferior advice that leads to poor decisions; many such specialists have insightful recommendations or know cutting-edge best practices. But when people pay a lot for advice, they become blind to its quality.

Francesca Gino, a professor of business administration at Harvard Business School and author of the decision-making book *Sidetracked*, has studied how people respond to advice and found that they're much more likely to follow advice when it's expensive than when it's cheap or free.[4] Gino conducted an experiment with university students in which half the subjects were randomly assigned to a cheap advice condition and half to an expensive advice condition, where advice cost twice as much. Gino determined the fee and each person decided whether he or she wanted to pay that much for an adviser. What happened once they received the advice? Participants who had paid for the expensive advice were more likely to follow it than those who had paid for the cheap advice. What's striking about her research is that people followed pricey advice even when they knew that it wasn't from a knowledgeable expert. Gino told them up front that the advice was coming from a peer, another university student who had just answered these same questions. She didn't promise it was good advice. She even showed them that the high price tag was arbitrary. When she was setting the fee, she essentially told each person: "We have this advice ready for you, but first we have to determine whether you'll get it for free or if you'll have to pay for it." Then she tossed a coin in the air while the person watched. So each student who had to pay a premium for the advice knew it was just his or her bad luck, nothing more. But once people knew the advice was expensive, they wanted to "get their money's worth," so they chose to follow the advice more often than the lucky people who had gotten the other side of the coin.

Keep in mind that these research participants didn't do anything to acquire that advice—they just showed up at a university building on campus, knocked on a door, and, voilà, advice was at the ready. But imagine what happens for those of us who have gone to great personal lengths to secure advice. What if your boss had spent several weeks researching and picking out the right consulting firm? Or what if your friend had recurrent, painful headaches and was paying two hundred dollars a week out of pocket and commuting an hour each way to see a practitioner who used hypnosis to relieve her symptoms? In both cases, there might be real benefits—your boss might get timely strategic advice, and your friend might feel relief for the first time in months. But because your boss and your friend are both investing a lot of time, money, and effort in getting the advice, they aren't going to be as savvy as usual in assessing it. And they'll certainly value that pricey, hard-to-acquire advice over anything you have to say. Your advice, even if you're on the boss's payroll, is usually seen as free.

What about inexpensive advice? Should you still be concerned if your boss has hired a professional consultant at a discount, perhaps someone who is just starting a business? Gino found that inexpensive advice is more persuasive than free advice (which means your boss still might find that outsider's input more valuable than yours), but the cheap stuff is still not as persuasive as the expensive advice. People expect to get what they pay for.

If you see that someone is planning to pay a lot for advice and you're concerned that he'll blindly follow it, sensible or not, there are several things you can do. Let's say your boss hires a consultant to offer advice on how to get more traffic on your company's website. You can be most helpful if you have input in the planning stages, before your boss is reading the consultant's advice and rationalizing why it's brilliant. Suggest to your boss that the consultant provide at least three top-level suggestions, not just one, so that your group can discuss those ideas and pick from among them. As we saw in the last chapter, research shows organizations make better decisions when they consider multiple, distinct options, but usually, the only option that's considered once a consultant gives advice is "Should we follow it or not?"[5] Another important step in the planning stages is to discuss under what conditions the company will decide to cut its losses and ignore the consultant's advice or set it aside for another time. If we triple our web traffic on our own in the next month, before we get the advice, should we just shelve the suggestions for a later date and keep doing what we're doing? Your boss may balk and say, "I'm paying a lot of money for this, I'm certainly not going to ignore it," but raising the question beforehand gets him thinking about how to evaluate the advice.

If you don't get a chance to have these conversations before advice is given, it's much harder to convince people to ignore advice they've invested in, but there are still some strategies you can try. Gino suggests that you ask your boss or your team to consider whether they would have taken the advice if it had been free. This can create some distance from the advice, though, depending on your boss's mood, you might be told where to go with your suggestion. Another strategy would be to ask your team to take the advice as a starting point, identify which idea they like most, and tailor it to the company's needs. That way, the boss still sees that the company is getting its money's worth from the consulting fee, but you ensure that your colleagues are thinking critically and not swallowing the advice whole.

Look at it from the other side of the table, from the consultant's perspective, and you'll realize this valuation is something you've probably known all along. Those who run consulting businesses know that setting a high price tag for services is key if they want to be valued. In his classic book *The Secrets of Consulting,* Gerald Weinberg sums it up in what he calls the "Second Law of Pricing": The more they pay you, the more they love you.

Getting Desperate

Imagine someone who has lost a lot of money gambling in Las Vegas. She has spent all the cash in her wallet and made several withdrawals from her checking account. This person could just walk away from the tables, frustrated and kicking herself but accepting that, yes, she lost all that money.

Some people might react this way and accept the loss, but many wouldn't. Instead, the desperate person might take out her cell phone and call friends, insisting she just needs to borrow two thousand dollars and she'll win it all back. She thinks about pawning a piece of jewelry or she tries to remember how much money she has in a savings account. All she can see is the escape hatch, the tiny chance that this loss isn't inevitable after all.

Now, some people will read this story and immediately picture a specific person, someone who has called and explained how everything will be better if he or she can just borrow some money. But many people will read this story and think, *Thankfully, I don't know anyone like that; I don't know anyone with a serious gambling problem.* But this isn't merely a gambler's addiction that nongamblers can self-righteously ignore. Most of us (and most loved ones we care about) have more in common with the person in this scenario than we'd like to believe.

What's key here isn't a gambling addiction, at least not for this book. What's key is that the person feels desperate, and desperation changes one's decisions. To see how desperation changes reasoning, try answering the following question:

PROBLEM 1: Which of these would you choose:
Get nine hundred dollars for sure or have a 90 percent chance to get a thousand dollars.

According to Daniel Kahneman, most people pick "get nine hundred dollars for sure."[6] It's delightful to think that someone would hand you that much money, and even if you didn't immediately think of a way you'd use that nine hundred dollars, you really like the idea of having it. You probably thought to yourself, *A thousand dollars is more money, but it's not guaranteed and I might come away with nothing. The risk isn't worth it.*

Now let's try a second question.

PROBLEM 2: Which of these would you choose:
Lose nine hundred dollars for sure or have a 90 percent chance to lose a thousand dollars.

You might have to think a bit more about this one, but most people would choose the 90 percent chance to lose a thousand dollars. Why are we willing to gamble here when we wanted the sure thing for problem 1? Because now we're backed into a corner, likely to lose money, and feeling a little desperate. That might not have been the word that popped into your mind, but that's the feeling. When individuals face certain loss looming over them, such as "Lose nine hundred dollars for sure," they feel a little desperate and quickly look around for some way out of the situation that might mean, fingers crossed, they won't suffer any kind of loss at all. Taking a gamble, which didn't look attractive earlier, starts to seem a lot more appealing.

This is based on a principle in economics called prospect theory that was first developed by Daniel Kahneman and Amos Tversky in the 1970s.[7] (Kahneman has a good sense of humor and admits that they deliberately chose a meaningless name for their theory—"Our thought was that in the unlikely event of the theory becoming important, having a distinctive name would be helpful in reducing confusion.")[8] These two scientists were trying to explain how people make financial decisions when they stand to win or lose money, but their work explains far more than money. It helps us understand motivations in many areas of life. Why does a desperate spouse continue to take back a partner who has walked out on her multiple times? There are many hopeful answers to that question: optimism, forgiveness, a belief that anyone can change. But given how people make decisions under desperate circumstances, this often abandoned spouse probably feels she can either accept a certain loss—the end of the marriage she's invested so much of her life in—or hope for the tiny chance that things might work out. As you saw with problem 2, when one option is a sure loss, that tiny chance of a win suddenly looks much more attractive. It seems very appealing even if it is a big gamble, even if it means, in this case, there's a strong likelihood she'll have to put up with him walking out again and deal with all the self-doubt and turmoil that follow.

"Desperate times call for desperate measures" should really be

"Desperate times delude and dupe us into taking desperate measures." When people feel they can't win, when they are confronted with almost certain loss, such as when they stand on the brink of bankruptcy, when they risk losing a license to do a job, or when they're about to be humiliated in front of peers, most of them don't stand tall and face the consequences. Most look for the trapdoor. They look for that sliver of light that might get them out of this dark place, regardless of where it potentially leads.

Of course, if we're the ones watching someone who's desperate, we see all of this differently. We're focused on the tiny odds of success. We're upset by how much it's going to cost her if it doesn't work out, and we're skeptical of her judgment every time she insists that it will work out. But the desperate person isn't motivated by how small her odds are. Her motivation lies in facing a choice between a 100 percent chance of losing and a 99 percent chance of losing. If she goes with the 99 percent chance of losing, at least there's a slim chance of winning, and that feels better than no chance at all.

When you think of someone in your life in a desperate situation, you might believe the problem is much bigger than prospect theory. For most people, desperate choices have dark histories. They might be in codependent relationships, where their choices feel bound to decisions of irresponsible loved ones. Maybe they have addictions to alcohol or prescription drugs, or maybe they're struggling with mental illness. Desperation doesn't usually pop up in a vacuum. But even if all those problems could be magically wiped away, a desperate situation still changes people and their perspectives. Take a person with exceptional mental and emotional health and drop him or her in a situation where imminent and almost unavoidable loss looms, and even the highest-functioning person you know is going to look long and hard for that trapdoor. When loss threatens, we all become risk-takers.

I wish that I had sure-fire advice for helping someone who feels desperate. But there are strategies worth trying. Since part of the problem is that the desperate person is framing the situation as a certain loss, look for ways to reframe the situation as a possible win. Where does she have something to gain? For a desperate friend who is thinking she'll do anything to make sure she doesn't lose her husband, perhaps you could gently frame the situation like this: "Which choice helps you show your children how women should be treated?" For a desperate colleague who is saying he'll do anything to make sure he doesn't lose his job, perhaps you could ask, "When you were younger, what did you want to be when you grew up?" Reminding a person of a larger set of goals can help him see an opportunity for change in his life, not just an imminent loss.

If reframing the situation doesn't work and you believe the person is going to take the tremendous risk anyway, consider helping to set up a tripwire, as discussed in the risk-taking chapter. If your desperate friend is going to take her husband back and try to make things work, what are three things she could watch for that would signal to her that she needs to talk with an attorney? If a colleague is going to offer to take a demotion and go part-time so that he can hold on to a fragment of his job, what future indications might tell him it's time to apply elsewhere?

Repeating the Same Bad Decisions

We often watch people make the same mistakes, or at least variations on the same mistakes, over and over again. Why does your best friend keep spending money each summer on camping gear when she never has time to camp? Why is your other friend always trying a new diet? Why does your boss repeatedly ask the least organized person on the team to lead presentations, and why does your brother-in-law think that his new idea for a business will be any more successful than his previous ones?

There are many possible reasons that human beings repeat their mistakes — everything from adamant denial to unrelenting optimism, from attempts to teach someone a lesson to attempts to redeem oneself. But scientists have recently identified two ways that people think about their pasts that distort decision-making and make all of us, not just the flaky types or Pollyannas, prone to repeat poor choices.

We generally think we will have more control in the future than we've had in the past. Two decision-making researchers, Elanor Williams at the University of California, San Diego, and Robyn LeBoeuf at Washington University in St. Louis, asked people how much control they believed they had in different situations.[9] Sometimes research participants were asked to think about something in the past, such as how well they'd done in a computerized game of chance they had just played or which movies they might have seen four months ago. Other times participants were told to imagine something in the future, such as how likely it was that they would win that same game of chance if they played another round this afternoon or what movies they would be likely to watch four months from then. You would think that people would use what had happened in the past as a gauge of the future. For the game the participants had just played, they might reason that they'd done a mediocre job, so the next time they played would be equally unimpressive. For the movies, they might realize they mostly watched high-suspense action thrillers, so that's probably what they would be picking again.

But that's not what happened. Williams and LeBoeuf found that people thought they would have much more control over themselves and over the world around them in the future than they had had in the past. When they had to predict how well they'd play a game, they thought they'd have a better chance of winning in the future. When they were picking movies, they said they'd be more likely to go to movies they felt they *should* see, critically acclaimed and sobering movies, rather than something they wanted to see. It's as though they thought *I've had sophomoric tastes in the past, but I'm on the verge of being classy.* They also thought they would have more control over their movie choices in the future than they'd had in the past. You can just imagine the defensive voice inside each subject's head: *My roommate picked all those action blockbusters. But in the future, four months from now? I'll pick something more serious, maybe something with subtitles.*[10]

At first glance, it looks as though they're just optimistic about their future, hopeful that they'll be more lucky or more intellectual. But Williams and LeBoeuf found that people also believed they would be more responsible for their future failures than they had been for their past failures. In one of their clever experiments, Williams and LeBoeuf asked half the participants to imagine that four months ago they had had trouble setting up a new DVD player, and the other half to imagine that four months from now, they would have trouble setting one up. When participants imagined a hypothetically frustrating evening in the past, a night spent futilely connecting wires, pressing buttons, and trying out unresponsive remotes, they said that the past experience wouldn't have been their fault. They said the instructions were probably poorly written or the DVD player was defective. But when other participants imagined that same hypothetical frustrating event happening in the future, they blamed themselves. They admitted they would probably be too impatient; they would leave the instructions in the plastic wrap, they wouldn't ask anyone for help, or they would skip a crucial step and not realize it. So it's not just that we're optimistic that our future selves will be better than our current selves; it's that we think we'll be in control in the future, whether things go well or poorly. But the past? *Why, I couldn't control that, not even in my imagination.*

This doesn't happen only in the lab — it's been documented in subway stations. A team of researchers led by Harvard University psychologist Daniel Gilbert approached people who had missed a subway train. The trains left roughly every ten minutes, and Gilbert's team stood around watching for commuters who'd just missed their morning trains. The enterprising researchers then went up to the people stuck on the subway platform and offered to pay them to answer some questions while they waited. When commuters had to imagine how they'd feel if they missed a future morning train by less than a minute, they said they'd feel a lot of regret and would be kicking themselves for getting up too late or stopping for a cup of coffee. They saw the future missed-train situation as something they could have avoided. But when the researchers asked them about the trains they had just missed, they didn't feel nearly that bad, and they certainly didn't feel much self-blame. When these latecomers reflected on the train they'd missed a minute ago, they blamed the world around them. "I wouldn't have missed the train if all of the gates had been open," "The machine wouldn't take my ticket," "There was this huge crowd at the bottom of the stairwell." When you imagine your future, even for something as mundane as catching the train, you think you'll have control. But as soon as you've lived through that moment and it becomes the past, you absolve yourself of responsibility in a snap.[11]

What does this mean for decision-making? Sadly, it suggests that when people look at a choice that turned out badly for them in the past, that poor decision doesn't seem relevant to the future, even if the circumstances are nearly identical. Dieters who have tried ten different weight-loss fads, none of which worked, eagerly buy the latest best-selling diet book. Or take your friend who decides to let her husband move back in, even though this decision has never worked out before. You're shaking your head because to you, these circumstances look disturbingly familiar. Nothing has changed in their relationship, at least nothing that suggests that this time will be any different. But your friend insists things have improved. She may point to a dozen different changes — how he's been promoted at work so he's feeling better about himself or how she's had less of a temper since she started doing yoga — but chances are what really makes this different in her mind is that she's comparing the past with the future. And whatever happened in the past, she didn't have nearly as much control over it as she believes she'll have in the future.

So how can you help? Once you recognize that your friend mistakenly believes she'll have more control in the future, there are a few things you can try. You can encourage what psychologists call "unpacking" the future, which has been shown to rein in a person's sense of control.[12] Ask her how the future *won't* be different from the past. She and her husband will likely still be living in the same cramped house they've owned for years. They will probably still have the same communication and relationship skills (or lack thereof), and they will certainly have the same parents and family history. Another way to help someone unpack the future is to ask why she won't have as much control over the future as she might like. Her company might be going through a reorganization, and with her husband's promotion, he doesn't know what his hours will be. Perhaps one of her parents is in bad health, stripping her world of any certainty. Make it clear that you have your friend's happiness and well-being in mind and then gently ask her to think out loud with you about those parts of the future that she can't foresee. I'm not saying that your friend will appreciate what you're saying in the moment. You're taking away control that she was counting on. But if you want to help her avoid another bad decision, trying this is much more useful than simply saying, "He's going to leave you again."[13]

Does this belief that we'll have more control tomorrow or next month explain the story at the start of this chapter? Did Lori's sister ignore her bleeding for so long because she kept thinking she'd have more control in the future? Perhaps—she might have thought she needed more fiber in her diet, and she could certainly control that—but there's another powerful explanation for her motives, and chances are it runs in all of our families.

Can't Turn Back

Most of us see ourselves as good people. You probably believe that you're intelligent, kind, and responsible and that you hum through life doing intelligent, kind, and responsible things.[14] You solve problems at work, you have integrity, you make good choices. What happens, however, when you do something that doesn't quite fit that view of yourself?

You're at work and your seventeen-year-old daughter sends you a frantic text—she didn't realize that her scholarship application had to be submitted by midnight. She has a swim meet and won't be home until after eight o'clock. Could you read over her essay and make sure it's okay? You say sure, and when you open the file, you notice some glaring errors. You start making tedious notes on a sheet of paper and realize it would be easier to edit the essay in Track Changes mode. Initially, you just fix the grammar, but by the end, you've also completely replaced one of her metaphors with something less clichéd. You send the file back to her, and she calls you the best mom ever.

But you're pretty uncomfortable with your choice. You might experience a pang of guilt, you might worry that she'll tell someone, but you'll also feel something that Leon Festinger, a social psychologist famous for his work at Stanford University, called cognitive dissonance.[15] Cognitive dissonance is the tension people feel when they are being internally inconsistent, the conflict between *I am honest and I've taught my children to be honest* and *I just wrote part of my daughter's essay.* Your thoughts clash with your actions, and the internal contradiction can be very troubling. You feel compelled to align your thoughts and actions so they're consistent again, but you can't undo your actions. You've already edited the essay and hit Send, so the only place with any give is your thinking. You don't want to see yourself as unprincipled, so instead you adjust your thinking: *She did 95 percent of the work and she would have caught those mistakes herself if she'd had more time.* You generate new thoughts so that you feel better about your decision: *She has strong grades and test scores — she has earned this scholarship on her own merits.* You feel less guilt, less tension. As long as all of this is true, you can still see yourself as an honest person. And if you never face that situation again, then your mental maneuvers helped you out of a tough spot (at least in your own mind).

The tricky part arrives the next time you face that decision, let's say a few weeks later, when her college applications are due. Now what do you do? She asks you to read over her essays again — your insights were so helpful the last time. Do you decide to edit her work because you've convinced yourself that's a fine decision, or do you change your mind and decide anew that it's wrong? If you change your mind, then you've contradicted yourself. And people hate to contradict themselves. So which way are you leaning? The choice that seems like less of a contradiction, the one that creates less conflict for you, is repeating your original decision and editing her work. Besides, your daughter has told you that some of her friends have professional coaches assisting with their applications. You can't afford that, so the least you can do is help her stay competitive. Of course integrity is still important, but you also need to do all you reasonably can to give your daughter the opportunity she deserves. By the time she sends you her third or fourth essay, instead of just editing a few sentences, you're reorganizing the whole piece to give it more punch.[16] The self-justification process contorts your reasoning and resets what it means to be a good person. You re-define *good* to include what you've done.

And that's very likely what happened for Lori's sister. That first time she saw the blood, she was very torn about what to do, but she decided, in the pressure of the moment, not to call the doctor. She probably felt cognitive dissonance, a strong tension between *I'm a smart woman* and *I just ignored that symptom,* but she generated a lot of thoughts that made that decision, just this once, acceptable. *There's no pain, it's the tiniest bit of blood,* and so on. When she faced that same situation a second time, a third, and a tenth, those same self-justifying thoughts, and perhaps several new ones, came to the rescue and she probably felt less tension each time, even as the symptoms grew worse. *What makes today's symptoms so bad that I need to call the doctor when I didn't call yesterday?* As outsiders, we feel more tension as we hear that she kept ignoring her symptoms, but the hardest decision for her was probably that first one.

In their book *Mistakes Were Made (but Not by Me),* Carol Tavris and Elliot Aronson provide a wonderful analogy that captures how the desire to reduce cognitive dissonance can lead a person to make the same bad decision repeatedly. They call it the pyramid of choice. Imagine a person standing at the top of a pyramid. Each side of the pyramid represents a choice, a choice he or she is making for the first time. Lori's sister had to choose between canceling her meeting at the library at the last minute or letting this symptom go for now. She could have gone either way. Once she decided to let it go for now, she began rolling down that side of the pyramid. She probably felt a need to justify why she hadn't made the other choice, why she hadn't prioritized her health. In fact, we know she thought long and hard about it at first because she shared her initial thinking with Lori, "There's no pain," and so on. Perhaps she didn't want to reschedule her meeting over something so embarrassing. All of these justifications convince her more and more that she made the right choice, and she moves away from that initial point of uncertainty, rolling a little farther down that pyramid of conviction. Each subsequent time she faces this choice and decides, *This really isn't the right time to call the doctor,* she rolls farther down that side of pyramid, until she's very far from feeling that initial point of indecision. The first time she made that bad choice, she could have gone either way, but before long, she has momentum behind her, and it's hard to make any decision other than the one she's been justifying to herself. Lots of small choices lead to one big commitment.

Secret bad decisions are especially susceptible to this kind of repetition. When you make a private decision that no one knows about, you might look around and try to figure out what others are doing, but the only model you actually have is your own. Others aren't talking about what they did when they faced this situation. No one is saying, "When it gets this bad, turn back."

So how do you help someone who feels that he can't turn back? (And men are just as susceptible to cognitive dissonance as women.) How do you redirect someone who has made a poor decision once and then feels compelled to keep making it? Part of the problem is that you don't always know when someone made the first poor choice. Lori's sister didn't tell anyone in her family about her symptoms for months.

But let's say that you do learn that someone has repeatedly made a poor choice, as Lori's sister did. The gut reaction is to exclaim, with simmering disbelief, "What were you thinking?" You lecture, you admonish, you browbeat. But according to social psychologists, these are some of the worst things you can do, at least if you want to see that person change his behavior and be up front about it. Remember that when someone keeps making the same poor choice, chances are he felt a lot of cognitive dissonance the first time he did it. He has been trying to see himself as an intelligent, competent person, and his poor choice threatened that view. When you ask him, "What were you thinking?" you're really saying, *Are you stupid?*[17] He feels threatened all over again. Chances are he will clam up and not share other choices he's made or is thinking about making. Rather than agreeing with you that it was a mistake, which he probably recognized right from the start, he feels the need to defend himself.

Keep in mind that when loved ones make the same bad choices multiple times, they are already finding it difficult to back down from their actions. Social psychologist Anthony Pratkanis, from the University of California, Santa Cruz, found that a much more useful approach than shaming is helping the person to feel supported and to recognize that he or she was trying to make a good choice.[18] In the case of Lori, she could have said to her sister, "I know you're smart and you normally take good care of yourself. Help me understand what your priority was when you first ignored your symptoms." Pratkanis points out that praising the person for having the right values will make it easier for her to talk about something that makes her feel incompetent. We all want to hear that we're good people, especially when we've made a big mistake or, worse, when we've made the same mistake repeatedly. As the psychologists Carol Tavris and Elliot Aronson write, when someone has been struggling with a string of terrible choices, it comes as sweet relief to hear someone else say, essentially, Yes, you are a decent, smart person, and you remain a decent, smart person, and that mistake remains a mistake. Now let's remedy it.

The Past Seems Perfect: Watching Your Parents Make Terrible Choices

It can be hard to watch anyone make a bad decision, but it's especially baffling and frustrating when that poor decision comes from someone who raised you. Your mom keeps spending hundreds of dollars on mail-order antiaging products. Your father finally has a buyer for his business, which means he can retire with dignity, but at the last minute, he adds a seemingly ridiculous clause to the deal, and his buyer backs away. Chances are that you were disabused of the notion that your parents were all-knowing by the time you were in high school, yet it still comes as a surprise, not to mention a dire omen, when you watch your aging parents entertain choices that you find highly questionable.

For most of this chapter, we've looked at the conditions in which anyone, young or old, might be prone to poor choices. But now let's look at one specific group: aging relatives. When your dad says he's excited to show you something he bought online, should you feel that dread welling in your stomach? Maybe, but science will show that you're probably feeling much more dread than necessary.

Perhaps the most important thing you need to know about aging and decision-making is that as people grow older, most of them experience something known as the positivity effect. Put simply, older adults prefer positive information over negative information.[19] You might be thinking, *Well, don't we all?* No one "prefers" bad news over good news. But this goes much deeper than simply wanting good news; it's more like wearing a big hood that partially obstructs your view of anything negative. Older adults can push the hood back and see past it if they genuinely want to, but normally, they're happy to keep the hood up, protecting them. As adults age, the hood increasingly hides the negative and guides their attention to the positive.

This positivity effect changes where we look — literally. Researchers have developed a clever apparatus that tracks where you're looking and records how long you fixate on different objects. In one classic experiment, a researcher sat people down in front of a computer that showed two faces. If one face looked afraid, adults under thirty spent a little bit longer scanning that frightened face than the other face, which looked happy, angry, or sad.[20] It's as though these younger adults were trying to figure out what had made the person afraid. But for older subjects, it wasn't the fearful face that drew them in — it was the happy one. Older adults spent more time looking at happy faces than at any of the faces expressing negative emotions.

This preference for the smiling faces doesn't affect just which television news programs your parents prefer. It also changes what details they pay attention to when they make a decision. In another study, scientists asked people to pick which kind of car they would hypothetically buy. To help them make informed choices, the scientists gave details about several different cars laid out in a grid. Compared to younger adults, older adults spent a greater proportion of their time reviewing the positive features of each car and less time viewing its limitations.[21]

When I first read about this study, I wondered if it pointed to a problem of failing eyesight and data overload. I couldn't imagine giving my grandmother a table of data about different cars — she used to struggle reading a long recipe in her own handwriting. But this focus on the positive happens even when the elderly can personally examine the object they're choosing. In another study, the researchers asked people of different ages to evaluate various objects, such as a key-chain flashlight, a ceramic coffee mug, and a clickable pen. They could turn on the flashlight, pick up the mug, write with the pen — basically do whatever they wanted to test each item. Compared to younger adults, older adults still paid more attention to the positive features of each object, and when asked which one they wanted to take home, they based their decision on the positive features, not the problematic ones.[22] If they liked white mugs, they picked the white mug, even if the handle was a little awkward or the mug a little too big for them. Mugs and pens may not be a big deal, but other researchers have shown that older people pay more attention to the positives when they are picking doctors and hospitals for themselves, and they are likely to overlook or wave away the problems with each choice.[23]

You can imagine the conversation playing out with your parents when they are shopping for a new car. "This car gets thirty-eight miles to the gallon," your mom points out in an enthusiastic voice. "Yes, Mom, but it's expensive to repair and it has a small trunk. You love a big trunk for groceries," you say. "That's true," she says, "but I do so love that red color." You grow frustrated because it feels like she's being overly simplistic, and she grows frustrated because it feels like you're being unreasonably negative.

If this filter obscured only what older adults noticed in the world around them, we could just make an extra effort to gently point out the negative features of this car or that computer when we're helping older people make purchases. But this filter also drapes across their internal world, playing with their memories. When older adults think back on earlier times in their lives, they typically remember more of the pleasant things that happened and fewer of the problems that occurred.[24] One of the most striking studies on the positivity effect in memory was done with a group of three hundred nuns. In 2001, Quinn Kennedy and Laura Carstensen, psychologists at Stanford University, along with Mara Mather (whose work on stress we saw in the previous chapter), asked a group of nuns to provide personal information about themselves, such as how often they exercised, any health problems they were having, and how often they felt various troubling emotions, such as anger, depression, and paranoia.[25] The nuns rated how they were currently doing in each of these areas. Then they had to think back and recall how they had been doing fourteen years earlier. You see, this wasn't the first time these nuns had answered these questions. Carstensen, one of the researchers on the project, had asked these same three hundred nuns the same questions back in 1987.[26] Because these nuns completed the same survey twice, the researchers could compare how the women had actually felt in the past with how they remembered feeling.

Naturally, the nuns forgot some things. They weren't exactly teenagers; the first time they picked up the pencil and paper for the survey, they ranged from thirty-three to eighty-eight years old, and the second time, the youngest was forty-seven and the oldest was a hundred and one. The researchers were especially interested in two groups — the youngest twenty-eight nuns (ages forty-seven to sixty-five in 2001) and the oldest twenty-eight nuns (ages seventy-nine to a hundred and one in 2001). All of the nuns, young and old, misremembered details of their lives when they looked back, but what is interesting is the consistent tilt in their memories. The oldest nuns in the group painted an overly rosy picture of their younger selves. They thought they had been happier and less paranoid fourteen years before than they had actually been. Women in their eighties and nineties also remembered having 35 percent fewer physical symptoms than they'd reported at the time. They misremembered or underestimated their physical ailments, thinking their bodies had felt so much better over a decade ago when, in truth, they had given the same number of physical symptoms at eighty as they had at ninety-four.

The younger nuns? Their memories didn't line up with the way they had described their lives the first time around either. The younger nuns had as many discrepancies in their memories as the older nuns, but the younger nuns' memories tilted the opposite way. The nuns in their fifties and sixties thought their lives fourteen years earlier had been *harder*, not easier. The younger women remembered being less happy than they had said they were at the time. They remembered being almost twice as depressed and twice as hostile as they said they'd felt, and they recalled struggling with almost three times the amount of paranoia. The younger women zoomed in on the negatives when they looked back on their lives, whereas the older women put a glowing spotlight on the positives.

One possible explanation is that the younger women had been less happy and had been having more health issues than they admitted when they'd filled out the first survey. Perhaps they felt pressured to sound happier and healthier than they really were, but fourteen years later, they felt comfortable being more candid about what they had been going through. Maybe, but then we need to explain why the oldest women showed the opposite pattern. We'd need to explain why the oldest nuns had claimed to be somewhat unhappy fourteen years earlier but now felt compelled to say, over a decade later, "You know, I actually felt great." A simpler explanation — and one that's consistent with many other research findings — is that once people reach a certain age, they start focusing on the positives.

This positivity effect has implications today for any of us with parents and grandparents who are still living on their own. When your older relatives are making decisions about their futures, you may be frustrated by how they view their past and use it to make predictions or assessments.[27] Your parents may say, "We had a wonderful time when we went to Orlando a few years back, so we're thinking about getting a time share down there." You're dumbfounded. You remember them calling you and complaining daily the last time they were in Florida, displeased by the heat, the crowds, and the bugs. You, being a much younger version of them, don't share that same positivity effect, and you wonder if this is an early sign of dementia.

Is the positivity bias a sign of mental decline? Researchers have been asking this question for the past decade, and the leading scientists on the issue believe that it is not. A recent neuroscientific study from the University Medical Center Hamburg-Eppendorf suggests that the ability to focus on the positives could actually be an indicator of strong emotional and mental health. Controlling one's emotions is no easy task, as anyone who's ever been cut off by a driver on the highway knows. It takes a considerable amount of cognitive resources and coordination among multiple brain areas to control emotions, so if an older person is successfully screening out the negatives, as frustrating as it may be to you, it might be good news for that person's mental health.[28] It's also helpful to realize that older adults *can* focus on the negatives, both in the world around them and in their own memory. They just ordinarily choose not to do so. One study found that when adults in their seventies were picking doctors for themselves or for someone their own age, they focused heavily on the positives of each option and largely ignored the negatives, but when they were picking doctors for younger people, they changed their focus and looked at more of the drawbacks, giving the pros and the cons of each doctor equal weight.[29] When they're thinking about someone who still has most of her life ahead of her, they're keenly aware that a physician's shortcomings matter.

Older adults even seem to have negative memories stored away and they can, if nudged just right, bring them to mind. In the study with the nuns, recall that the oldest women painted an overly rosy picture of the past, making it seem as though the negative memories had eroded and the happy memories were all that were left. But researchers established that the oldest women were able to recall the problems they had been having fourteen years earlier, although only when told, "It's important that you're as accurate as possible."[30] Once prompted in this manner, the older women dredged up the less pleasant memories.

What's going on here? Why wouldn't they try to be as accurate as possible all the time? Laura Carstensen and Mara Mather, two of the lead researchers, believe that older adults are trying to control their moods.[31] Increasingly aware that their time is limited, the elderly are less inclined to focus on the problems in their lives and more motivated to remember the past and see the present in a way that's emotionally soothing or comforting. They want to be satisfied with the lives they've lived and with the options they still have available, so they focus on the good things. This also explains how your aging parents can tune out the problems with their choices but still spot the faults in the options you're considering—they are trying to improve how they see their own lives, not how they see yours. They want to be content with the lives they've chosen, with the spouses they've picked and the money they've spent, but for you, they want something even better.

Thankfully, many adults develop the skills they need to keep their moods in check. Carstensen and her colleagues found that most people, men and women alike, became more skilled at regulating their emotions over the course of their lives.[32] They're not always successful, of course—your mother may still lose her temper, and your father may have times when he's deeply sad about something he won't discuss. But in general, as people age, they become better able to choose their moods, and the moods they choose tend to be pleasant ones.

This is good news. You don't have to be concerned about leaving your children with your parents for a week—just because your parents seem overly focused on the positives in their own lives doesn't mean they're blind to negative consequences for your kids. They might still let your kids have more doughnuts than you would, but that's just grandparenting.

There's another important way that older adults differ from younger adults when they make decisions: they usually want fewer options. Studies have found that as people age, they seek less information when they're making a decision and are happier with fewer alternatives.[33] Researchers at Cornell University asked older adults and younger adults how many options they'd ideally like to have when they're making choices. They asked about everything from jams ("Would you like two kinds of jams on the breakfast table or six?") to apartments to prescription plans, and, on average, older adults wanted about half as many options as younger adults.[34] Older women even wanted fewer options for treating breast cancer than younger women. It's not that choices don't matter to people once they reach sixty-five, but a grandmother who wants a say in her breast cancer treatment might not be remotely interested in eight months of chemotherapy. Knowing that older adults prefer fewer options could be valuable knowledge when you're discussing an important decision with your parents or with someone thirty years older than you at work. It might seem as though an older person isn't interested with the decision process or isn't giving enough attention to all of the options, but she might be much more engaged if you simply eliminated a few of the least attractive alternatives.

At what point do these changes start to occur? For the positivity effect, the answer seems to be "as soon as you graduate from college." Thirty-year-olds have a more positive tilt to their thinking and memories than twenty-year-olds, and at forty, the world looks even better to you than it did at thirty. One research team that has tracked the positivity effect says that most people will see the world at its rosiest in their mid- to late sixties and then, as long as they don't suffer a debilitating health problem, will enjoy that mindset for a while.[35] Personally, I find that the oldest nuns give me hope. Even in their nineties, they still had delightful memories of their eighty-year-old selves.

Every individual is different, of course, and your grandmother may complain continuously that she doesn't have enough options and be frustrated with everything, from the limited alternatives her doctor offers for managing her arthritis to the selection of beverages at Denny's. But most people become less preoccupied, not more, with the frustrations in their lives as they age. This runs counter to the image of Statler and Waldorf hurling insults from the *Muppet Show* balcony, and, yes, some old men do sit and complain about the world in front of them. But gliding contentedly through one's golden years is much more common.

Strategies for Helping Your Parents and Older Relatives

If there's a really important decision that your parents or grandparents have to make and you're concerned that they won't choose at all or will choose poorly, is there anything you can do? Help them narrow their options. They want to find a new apartment so they can downsize? Don't send them thirty listings, even though you might want to review that many, if not more, if it were your move. Find out their most important criteria and then send them five or six listings that match what they're looking for. Don't assume that a big decision means they will want more options. All of us, young and old, prefer more options when the choice is trivial and fewer when the choice is really important. When you walk into a gas station, you're glad to see that there are a hundred different types of candy — that's fun. But if the pumps outside had that many types of gasoline and half of them would ruin your engine, you wouldn't spend time choosing — you'd drive away.

What if the positivity bias is the problem? Let's say that your parents are leaning toward a particular apartment and there are obvious red flags that don't seem to concern them. You're alarmed about the rent or the fact that it has a lot of stairs. One way to tackle the positivity bias is to ask them what they would suggest for another person with similar needs — someone they respect; a cousin, say. Older adults are better at taking the negatives into account when they consider whether the choice is good for someone else. Of course, they may emphasize how they are healthier or wealthier than the cousin, but at least you've encouraged them to admit that, yes, there are problems with that apartment.

Most of us would probably approach an important decision when we have everyone's full attention. In helping your parents pick an apartment, you might be tempted to raise the topic over dinner some night or to sit them down and say, "I want to be sure you make the best choice." Though this strategy might seem sensible and persuasive, a more effective tactic would be to raise the topic when they're distracted. Bring it up while you're making dinner with your dad or while you're cleaning out the garage with your mom. This isn't an attempt to trick them into a decision they'll regret later; it's an attempt to get them to see the negatives of the option they're considering. Research shows that older adults become overly focused on the positives when they give a decision their full attention.[36] Focusing on the positives doesn't happen automatically. It takes effort and concentration to regulate one's emotions, to keep oneself focused on the positives, so if something else is competing for your parents' attention, say a pot of pasta or a stack of boxes, they're more likely to see the negatives for themselves without your having to point them out. Be warned — they may also become annoyed and irritated with you for bringing up the decision when they've so successfully managed to ignore the negatives up until now. But at least you'll know they're aware of the problems.

It may also help to realize that just as your parents probably have a positive bias, you probably have a negative one. Younger adults, especially in their late teens and throughout their twenties, tend to scrutinize the downsides of the options they are given. That doesn't mean younger people are worried or pessimistic, just that they weigh their concerns heavily when they're making decisions. You don't have to admit to your parents that you have a negative bias. But admitting it to yourself might make it easier to accept that their reasoning is a lot different from yours.

I talked with Lori again almost a year after our initial interview. Things turned out well for her sister. She had surgery and then underwent several months of chemotherapy, and the scans showed the cancer was gone. She was in remission. When I asked Lori if she still wondered why her sister kept delaying the decision to see the doctor, she said, "You know, my brother and I haven't talked about it in a long time. I still don't understand why she did that, why someone with her common sense would wait so long. But what can you do?"

I hope that if she reads this chapter, she'll be able to help her friends see what they could do in the same situation.

Chapter 6 at a Glance:
Watching Other People Make Terrible Decisions

THINGS TO REMEMBER

1. Just as poor driving conditions make it hard to drive, poor decision conditions make it difficult to decide.
2. Women find themselves poised to play decision-support roles more often than men.
3. Beware of people who have paid a lot for advice.
   - A high price tag blinds people to the quality of advice they've received.
   - Example: Boss hires a consultant
4. Desperate times delude us into thinking we have to take desperate measures.
   - Prospect theory says when you face a certain loss, you become a risk-taker.
   - Example: Would you rather lose nine hundred dollars or have a 90 percent chance of losing a thousand dollars?
5. Human beings often repeat bad decisions (fad diets, new business ventures, taking back cheating husbands).
   - We think we'll have more control in the future than we've had in the past.

- Examples: DVD player that won't work and subway riders
6. Attempts to reduce cognitive dissonance lead people to repeat mistakes.
    - Example: Helping your teenager with a college application essay
    - We redefine *good* to include what we've done.
7. The positivity effect means older adults prefer positive information.
    - Example: Picking out a car to buy or a mug to take home
    - Nun study: Older nuns remember their lives being better than they actually were, whereas younger nuns remember their lives being worse.
    - Older adults focus on the positive to regulate their moods.
8. Older adults prefer fewer options when they're making decisions.

THINGS TO DO

1. Make a plan for how you'll evaluate advice before you get it.
2. To ensure that you won't be lured into regrettable actions when you're feeling desperate, try reframing a certain loss as a potential win or setting a tripwire.
3. If you want to make sure your parents have considered the negatives of a choice, raise the topic when they're slightly distracted to reduce the positivity effect.
4. If you're frustrated with a decision someone has made, instead of asking, "What were you thinking?," say, "You're smart. Help me understand."

Afterword

REGARDLESS OF WHETHER you're a woman or a man who works with women, I hope you've gained something of value from this book. I hope you have a few more nuanced strategies to use the next time you're tempted to go with your gut, the next time anxiety is making it impossible to decide, or the next time you wish one of your coworkers would show a little more confidence in meetings. Decision-making is hard for all of us. But using this book ought to make it easier.

We've looked at several ways to make sure you take appropriate steps when making decisions of consequence. You can think of confidence as your dial, turning it down when you're still gathering information and making your decision but turning it up, way up, when you need followers. You can set a tripwire so you remember to reevaluate a reversible choice in the future; if you discover you've made a second-rate decision, you won't live with it longer than you need to. And if anxiety is your nemesis, making it hard to step onto a particular path, now you know to tell yourself: *This isn't anxiety. This is excitement.*

But perhaps the most important message in this book focuses on how we see one another as decision-makers. Women have more exceptional judgment than society would lead us to believe. Women tend to rely on conscious, data-driven analysis just as often as, sometimes more than, men. And a data-driven choice is an informed one. We've seen that many people believe that "women take care while men take charge," but now we can stop misinterpreting a desire to collaborate as an inability to decide. And we can begin to see women, especially women professionals who have spent years building their expertise, for the effective risk-takers that they are. If we can start recognizing that her judgment is just as good as his, then it won't feel like such a risky bet to put women in top leadership roles.

So what's my parting piece of advice for anyone who wants to keep learning how to make better decisions? I have one more story for you. I'd like you to meet Zoe.

Zoe is the kind of person who refuses to fit into a tidy little box. The daughter of working-class parents, she attended a women's college, spent more time with her books than her friends while she was there, and graduated magna cum laude in the top 2 percent of her class. If you guessed that Zoe is intelligent, thoughtful, and serious, you'd be right. She also wears heavy black eyeliner, dyes her long hair either Marilyn Monroe blond or hot pink, and plays the bass guitar in an all-woman punk rock band. During the day, she manages a program to help other young adults forge a career path and find their life calling. But at night, she's found hers. When Zoe steps onto the stage and tunes her guitar, when she starts sweating and bobbing her head under the blue strobe lights, she couldn't be happier.

The first time I interviewed Zoe, it was May 2014, and she was in eastern Colorado, eager but exhausted. She was sitting at the back of her band's bus, separated from the other band members by a wall of sleeping bags, suitcases, and sound equipment; the band was making its first coast-to-coast tour. The others had slept in and gotten a late start that morning, but they were making good time on the highway. They'd had nine consecutive nights of enthusiastic, wall-to-wall crowds from Rhode Island to Missouri, and they'd be headlining when they pulled into Denver. Her success was in full bloom.

She didn't start out in music with the hope of someday being on tour. Three years earlier, back in 2011, she had asked some friends if they wanted to form a band. Zoe had been intensely studious through-out college, and after graduating she recognized a need for more fun in her life. Two of the four women who joined the band didn't even know how to play instruments, but they had an experienced drummer and figured they'd learn together. They recorded a couple of tapes, then moved up to an EP. They wrote their own songs and, before they knew it, they were booking their own shows. Their sudden success was so unplanned they had to rent a van each time they drove to a distant gig.

In 2013, NPR featured their music on World Café. The host de-scribed the band's first full release as "one of the best rock albums of the year," and doors began to open. The band was written up in the *New York Times, Pitchfork, Spin,* and *Elle.*

Several months before that van ride to Colorado, Zoe had to decide how much she was willing to commit to the group. Up until that point, she'd been doing what she'd set out to do — simply having fun. She could hold a full-time day job and play with the band on nights and weekends. The band had never hired an outside person — there was no attorney, no booking agent, no manager. They'd played DIY concerts up and down the East Coast, but if they wanted to make money, they would need to go national, make it all the way to stages in California and Texas. She'd have to take more than six weeks off from her job, forgoing the paycheck that covered her rent and her bills. She'd be turning her whole life, at least for a few months, toward the band.

The band's incredible budding success gave her the confidence to make the commitment to go on that fourteen-thousand-mile, thirty-concert tour. "Once I saw how much people liked our music, it was easy to keep going. There was a ripple effect, and with every small step, it got easier to take bigger steps," Zoe said. She didn't start off wanting the band to be huge. Zoe calls that kind of thinking "premeditated disappointment." She just kept deliberately asking herself whether she was willing to take the next step.

After our first conversation, I connected with Zoe again in September of 2014. Their six-week cross-country tour had been such a success that the band confidently went back out on the road again, easily booking big venues. This time, things soured. One of the women—their lead singer and guitarist—quit the band abruptly. They hadn't found a replacement yet, and since they drove to a new city every afternoon, their chances of finding someone were slim. "It's not a great tour," Zoe admitted. "I'm disappointed."

Not surprisingly, the tone of this conversation was different. "With the band shakeup," she said, "it's made me think, for the first time in a while, what if this isn't worth it? What if the band isn't the good long-term path that I wanted it to be?" I could understand Zoe's thinking. All of us reevaluate decisions when events take an unfortunate turn. But she also said something I didn't expect. "I think I would have been happy with graduate school," she said.

Ah, graduate school. Zoe was reaching back to a decision she'd made four years earlier, a choice she'd made long before she'd even considered starting a band. She'd received a hefty scholarship from her alma mater that she could allocate to any PhD program to which she was accepted. Her undergrad professors were convinced she could get into the best doctoral programs in the country, and Zoe dutifully spent months pinpointing programs that fit her interests and that would be within driving distance of her parents. But she also took several analytical steps that most people don't when they consider graduate school: she researched the current and expected job market for PhDs in her field, traced the career paths of the graduates from the programs she was considering, crunched the numbers on how much debt she'd rack up (even with a scholarship), and read firsthand accounts of graduate students who felt overworked and undersocialized. What she saw didn't fit her vision of the good life. Based on all of her research, she knew with incredible clarity that she didn't want to go to graduate school, at least not at that point in her life. In an unprecedented move, she gave the scholarship money back.

Zoe had told me about that graduate-school decision the first time we talked, when she was on the highway to the Denver show. When she'd described it then, she was happy with that hard decision. She remembered feeling relieved when she decided not to use the scholarship. And, personally, I was impressed, not only with her bravery for saying no to what could have been an obvious choice, but with how much research and analysis she'd put into her decision. Of my friends who have considered graduate school, I don't know anyone who has done that kind of homework. Even Zoe's undergraduate professors, the ones who had nominated her for the scholarship, applauded her for her decision when she explained it to them. More than one of them admitted that Zoe had tabulated hard realities they hadn't even considered when they'd chosen to go to graduate school.

So why did Zoe later doubt that decision? She had felt confident about that decision for three and a half years after making it, and it had been the right choice for the right reasons, and none of those reasons had changed. The job market for her field hadn't improved. Grad school was still grueling. And it wasn't as though she'd passed up grad school because she had set her sights on being a rock star — she hadn't even thought of starting a band at the time. Just because the band might dissolve, just because the decision to commit to the band might have been shortsighted, didn't discount the wisdom of her graduate-school decision.

This wasn't the first time I ran into this pattern. A number of women I interviewed for this book had unexpected levels of success, and when opportunities they hadn't imagined were opening up left and right, they were pleased with their past decisions and valued their decision-making process. They looked back on those times when they'd taken risks and invested their whole selves, and they were highly confident that they'd chosen well. It spurred them to risk again. But if I spoke with any of them on another occasion, when a decision looked like it might have a bad outcome, the person looking back didn't question just that one, isolated judgment. She sometimes questioned her ability to make any judgment at all. After a setback, each went through her Rolodex of big decisions and noted how flawed every choice was. One upper-level manager who was regretting her decision to change jobs a year after she'd moved across the country began questioning all kinds of choices, decisions she'd made as a parent, a homeowner, and a wife. "It all comes tumbling down, doesn't it?" she said.

It's as though we keep judging our past decisions by the latest developments. Not by the process that led us to our choices, not by all the excruciating pieces of analysis we've done or even by the ways our lives measurably improved following that decision. All too easily, we judge our decisions according to how today's afternoon light falls on yesterday's choice, making it hard to say with any conviction, "That decision was right for me."

My male friends are quick to point out that this pattern isn't unique to women, that we all reinterpret our past decisions from today's vantage point. I believe that. But I'm also convinced it's not the same. If the only voices of doubt a woman heard were her own, that would be one thing. But a woman's voice is really adding to the chorus of questioners willing and ready to judge her choices. As we've seen throughout this book, society says women's decisions need to be scrutinized more than men's, especially when those decisions go badly.

The Misleading Harmony of Memory

We're convinced we'll remember how we thought and felt about decisions we've struggled through. How could we forget? Especially for the big decisions we've made, the big transitions in our lives that consume so much of our time and energy, it seems as though we'll always remember the experience. Those decisions seem perfectly preserved in our own personal memory vaults.

But they're not. Memories are malleable. As one famous memory researcher, Elizabeth Loftus, puts it, memory is not "like a museum piece sitting in a display case."[1] You don't just walk through and admire your memories from a respectful distance. Every time you access them, you reconstruct them, messing with them in the process. Instead of an art museum, envision an art class. Imagine that an art instructor tells a class to go observe cows and then come back and paint pictures of those cows.[2] Each person's painting will look different from everyone else's, and if a student is asked to paint that same cow and pasture again, he or she will paint a slightly (or potentially very) different picture than the first one.

And just like your second or third or tenth rendition of a painting, memories don't always improve. Your most recent painting isn't necessarily your favorite. But whatever you're remembering, it will probably reflect your current outlook. You keep reworking your memories so they better match what you're thinking and feeling now, whether that's better or worse.

When parents of toddlers try to remember three years back, reconstructing how they reacted to their baby's crying, they're likely to misremember their actions. But their memories aren't random. They err on the side of whatever experts recommend today, whatever child-rearing philosophy the parents subscribe to currently. If experts say it's better to let the baby cry it out, parents are likely to remember having sat in another room waiting it out more often than they actually reported at the time. If experts say it's better to pick the baby up immediately, parents remember having jumped up at the slightest peep more often than they actually did. We could chalk that up to sleep deprivation; new parents don't sleep much and no one's blaming them for a year of sketchy memories. But other people update their memories to fit their current outlook as well. College students revise their memories of their past grades to fit how well they're currently doing in school; voters update their memories of how much they liked or disliked a candidate after an election; and thirty-year-olds edit their memories of their teenage personalities to better match their adult selves.[3] We update our memories so they fit what we're thinking and feeling now. We keep changing our story. As social psychologists Carol Tavris and Elliot Aronson put it, people bring "their past selves into harmony with their present ones."[4]

This helps explain why, when you're feeling optimistic about your life, when things are better than you'd ever imagined they could be, your past decisions look insightful, even fated. That halo glow over your judgment gives you confidence to approach the next choice on the table. But when you're feeling hopeless, those past decisions look flawed, and so do you and your reasoning process. In those moments of doubt, it's easy to be critical of the decisions where you clearly took a risk. If you're Zoe and the lead guitarist has quit mid-tour, you find yourself questioning a decision you made long before you'd even met that guitarist.

The Handiest Decision Guide You'll Ever Meet

As we've seen throughout this book, many of the subtle and not-so-subtle messages we hear about women's decisions aren't true, but women's ability to make smart choices can be threatened by those messages.

My last piece of advice applies to men and women alike, but it's especially important for women. It's one more strategy for ensuring that women keep those affirming messages coming.

Women need to become historians of their own decisions. They shouldn't be rewriting their decisions with every change in circumstances. So my last piece of advice? Stay true to your story. Begin capturing snapshots of your decision process so that when you look back, when your circumstances and perspective change, you can more accurately reconstruct what you were thinking and feeling at that point. This will help you remember what mattered most at the time, the wisdom that you had at that juncture. It also helps you learn from your life.

With technology, there are probably dozens of creative ways to capture your decision process. I'm going to suggest one approach, and it's low-tech: Start keeping a daily one-sentence journal. The notion of writing in a journal every day will sound time-consuming and unrealistic to some — you might be protesting, "I can't even find a clean pair of socks every day" — but hear me out. You don't need to carve out twenty minutes each evening to keep a journal and you don't need to clear your desk or rearrange your life. Just write a sentence or two each day. It takes five minutes, tops, especially if you make it part of your daily routine, jotting down a note or two before you go to bed.

How is writing a few lines a day going to help you track your decisions over time? Don't write the sentence just anywhere. Buy a one-sentence journal. You can find them online or at a bookstore. A one-sentence journal is organized with 365 pages, one page for every day of the year, but each page typically has space for five entries, one for this year, one for next, and so on. These are just snapshots, really, a sentence or two each day.

With a regular journal, you write an entry and you turn the page, always moving forward, rarely moving back. You don't review your past thoughts, feelings, or decisions unless you intentionally make the effort and go searching through for a particular entry, which is time-consuming and not something most of us think of or want to do. With a one-sentence journal, the structure allows you to see those snapshots. It's like a photo album of your decisions. The first time through the journal, you're recording your thoughts and experiences on a new page each day, but the next year, you'll have a chance to revisit those thoughts and feelings when you capture that day's events. The value of the journal grows with time as you revisit how you really decided.

What do you write about? Write a sentence or two about how you were feeling that day and what you really valued. What went well and what went poorly? When you made a decision, what was the most important thing, or what initially held you back?

Consider the two entries below, taken from my own journal. (I see now that they're actually three sentences, but short ones.)

April 11, 2013 I did it—I broke down and got a comfy, ergonomic desk chair. Not sure why I waited so long—the expense? Improving where I spend 7 or 8 hours a day is money well spent.

August 17, 2013 So good to be home. I love my first day back from a trip, especially if I can spend it at home. I eat my favorite foods, walk the neighborhood, love the feel of my life.

Perhaps all you need to do is make an observation about your life once, and after you have that spark of insight, every decision you make from that day forward reflects that helpful, added lesson about yourself. Not me. I write it down or talk about it with my husband and feel pleased with my newfound self-awareness, but the next time I consider an indulgent office purchase or a colleague wants to schedule a meeting the day after a trip, I fall into my routine. I lose sight of what made me happy or frustrated in the past, and I forget the promises I made to decide differently in the future.

But I'm more likely to remember with a daily journal. For both of these entries, when I was writing on these same pages a year later, I scribbled down something like *So true — good insight last year*. And because I've found a way to remind myself, I make better choices. I've continued investing in my office (I recently bought a window air conditioner), and whenever possible, I keep the day after a big trip open.

I've kept a daily one-sentence journal since 2012 and it's corrected many of my own assumptions about my decision-making. Confidence is probably where I've undergone the most change. I've noticed that sometimes I'm very confident and purposeful about a decision I'm making and sometimes I feel like a hot mess. But when a decision turns out to have been a good one, I'm surprised to discover later how much doubt I felt, how unclear the right choice actually was at the time. For good decisions, the doubt washes away. That's a valuable realization because now, if I'm not confident of a decision, I know that's not a harbinger of doom. Now that I no longer expect a resounding click of confidence to precede every smart choice, I find it easier to decide. Not easy, but easier.

Note to Future Self

The first time I interviewed Zoe, she told me, "The decision not to use that scholarship was the bravest, most adult decision I'd ever made. I didn't know what I wanted to do, and that was clearly the wrong way to start graduate school." Would it have helped Zoe to read those words years later? Would it have soothed her doubt to hear how courageous she felt when she'd made the decision? I don't know. But I can't help noticing how the timing of the two key events lined up. She made her decision to forgo graduate school in September 2010. Exactly four years later, in September 2014, her band's lead singer quit and she faced her crisis of confidence. If she'd captured her initial brave decision, if she'd described her presence of mind surrounding her hard scholarship decision, she would have been reading those notes to herself just as she needed to hear them.

There's no guarantee you'll have perfect timing, that a message you need to hear will come exactly on the day you need it. But you'll have a system for capturing your memories and combing through them, for reminding yourself what led you to this choice and away from that one.

We women need to send messages to ourselves. The world sends a lot of doubt our way and we can counter it with our own understanding and careful thinking. We often have better judgment than we realize, than we remember, and we just need a little reminder from ourselves of how wise — and sometimes brave — we are.

Recommendations for Further Reading

If you're looking to read more about women and leadership, try

- *Lean In: Women, Work, and the Will to Lead,* by Sheryl Sandberg (New York: Knopf, 2013). This book has fueled a global conversation about women in the workplace. If your colleagues have read only one book about gender, chances are this is the one.
- *Mistakes I Made at Work: 25 Influential Women Reflect on What They Got Out of Getting It Wrong,* by Jessica Bacal (New York: Plume, 2014). The author interviewed successful women in different professions, from writers to entrepreneurs to food critics. These women described mistakes they made in their careers, both big and small, and how they became smarter, better leaders because of those missteps.
- *What Works for Women at Work: Four Patterns Working Women Need to Know,* by Joan Williams and Rachel Dempsey (New York: New York University Press, 2014). A well-researched book that identifies four biases that many women face in their careers. The authors offer a variety of concrete strategies for tackling these biases when women encounter them.

- *Daring Greatly: How the Courage to Be Vulnerable Transforms the Way We Live, Love, Parent, and Lead,* by Brené Brown (New York: Penguin, 2012). Although you'll probably find this in the self-help section of your bookstore, it's a powerful book for all people who want to lead with their whole selves. The core message is "Vulnerability is not weakness."
- *How Remarkable Women Lead: The Breakthrough Model for Work and Life,* by Joanna Barsh and Susie Cranston (New York: Crown Business, 2009). This book provides compelling reasons why both men and women should embrace leadership styles that have traditionally been labeled feminine.

If you want to read more about decision-making and biases that surreptitiously shape our choices, pick up

- *Decisive: How to Make Better Choices in Life and Work,* by Chip Heath and Dan Heath (New York: Crown Business, 2013). If a friend wanted to learn how to make better decisions, this is the book I'd recommend. It provides clear advice and fascinating research in a friendly, accessible style.
- *Mistakes Were Made (but Not by* Me): *Why We Justify Foolish Beliefs, Bad Decisions, and Hurtful Acts,* rev. ed., by Carol Tavris and Elliot Aronson (Boston: Houghton Mifflin Harcourt, 2015). Want to see how we're all in the same boat of self-justification? Read Tavris and Aronson and discover how hard you work to twist your mistakes into wise choices.
- *Predictably Irrational: The Hidden Forces That Shape Our Decisions* (New York: HarperCollins, 2009) and *The Upside of Irrationality: The Unexpected Benefits of Defying Logic* (New York: HarperCollins 2010), both by Dan Ariely. In these two books, the author takes us on an entertaining and clever tour of human be-

havior where the guiding questions are "Why don't we make rational decisions?" and "Could there be a silver lining to our irrational ways?"

- *Thinking, Fast and Slow,* by Daniel Kahneman (New York: Farrar, Straus and Giroux, 2011). The author explores how there are two systems that drive the way we think, one system that's fast, emotional, and almost impossible to control, and one that's slow, analytical, and thorough. We'd like to think that the second system is in charge, but all too often, it's the first system that leads and gets us into trouble.

- *The Invisible Gorilla: And Other Ways Our Intuitions Deceive Us,* by Chris Chabris and Daniel Simons (New York: Crown, 2010). Another thoughtful book on how we fool ourselves. Chabris and Simons peel away our convenient but all too faulty assumptions about memory, attention, and reasoning and show that it takes a bit of work, but you can be a lot smarter than you were yesterday.

- *Whistling Vivaldi: How Stereotypes Affect Us and What We Can Do,* by Claude Steele (New York: W. W. Norton, 2010). Like it or not, we're all subject to unconscious biases, stereotypes that shape how we see ourselves and others. The author shows us how our notions about race and gender unconsciously disrupt our behavior.

- *Blindspot: Hidden Biases of Good People,* by Mahzarin Banaji and Anthony Greenwald (New York: Delacorte, 2013). If you've ever wondered why we are so quick to judge others or why good people discriminate, you'll find answers in this book. It includes helpful self-tests to gauge where your own biases lie.

If you like understanding how people are similar and different, consider

- *Quiet: The Power of Introverts in a World That Can't Stop Talking,* by Susan Cain (New York: Crown, 2012). This last book doesn't fit neatly in either of the categories above, but it's one of the best works on the similarities and differences between two groups, introverts and extroverts. It's thoughtful and thought-filled.

Acknowledgments

If you want to become a successful novelist, "you need three things," says Michael Chabon, "talent, luck, and discipline."[1] If you want to become a successful nonfiction writer, I believe you need a fourth thing: a team. A team of smart, hard-working people who persevere through your sprawling drafts, encourage you to come up with crisper arguments, and pass along the juicy articles they read over breakfast.

And I've been lucky enough to have an incredible team.

My agent is Lindsay Edgecombe, and if you're of the mindset that literary agents aren't needed anymore, you haven't met Lindsay. She's smart and skilled at giving feedback, and when I can't find my point, Lindsay finds it for me. I'm the anxious type; Lindsay sees that, and she still works with me. I am grateful to the entire skilled team at Levine Greenberg Rostan who helped this book see the light of day, including Jim Levine, Beth Fisher, and Kerry Sparks.

I've had a chance to work with not one but two sagacious editors, Courtney Young and Jenna Johnson. Courtney lobbied hard to acquire this book for Houghton Mifflin Harcourt, and she read the initial, all-too-tedious drafts, nudging me to drop the social-scientist voice and say things that were truly useful. When Jenna joined the project, she said, "You lost me here," in all the right places, but she also said, "I love this," regularly enough to keep me motivated. If you find yourself quoting this book, thank Jenna.

I encountered many good people at Houghton Mifflin Harcourt. Pilar Garcia-Brown helped me stay on top of deadlines and connected me with talented people. Tracy Roe made sure that if I started a sentence in the past tense, I ended it that way. Tracy also went beyond the call of duty as a copyeditor, and if I ever need an urgent-care physician in Virginia, I'll know where to go. Martha Kennedy designed the book's fantastic cover, Loren Isenberg conducted a careful legal review, and Kimberly Kiefer managed the entire production. Rachael DeShano and Emily Andrukaitis oversaw the editing process and made deadlines a pleasant affair.

On the marketing and publicity side, Kathleen Marie Perkins has been my marketing guardian angel, sent straight from heaven to get me over my self-promotion hang-ups. If you've heard my voice in an interview or read an article about my work, it's because of the brilliance and persistence of not one or two or even three people, but seven: Lori Glazer, Laura Gianino, Giuliana Fritz, Ayesha Mirza, and Carla Gray at Houghton Mifflin Harcourt, as well as Gretchen Crary and Jessica Fitzpatrick at February Media. Gretchen, a special shout-out for holding my hand as I took what felt like bold steps into social media. Kayleigh Shawn McCollum made me forget that she was taking photographs, and Alyssa Wang encouraged me to think bigger than "I'd like a website that I can't break."

Dan Simons introduced me to my agents, played the skeptic when I needed one, and gave me the idea to write this book in the first place. If you're not familiar with Dan's work, please set this down and Google his gorilla video. Dan makes us all smarter.

Thirty-four women were generous enough to allow me to interview them and reflective enough to walk me through some of their toughest decisions. Their stories were powerful, and they gave me not only women to write about, but women to write for. Another two dozen women participated in the Decision Dialogues that I held in the early research phase of this project. I wish I could thank all of these women by name, but I promised to lock their real identities away so that they could tell their stories candidly. To all of these women, thank you.

A scientific work like this requires tons of research. Nicole Brous proved to be an extraordinary research assistant, tackling both the tasks I asked her to and the ones she saw I wasn't completing on my own. Theresa Earenfight offered inspiring stories of women in history, and Carole Levin answered some timely questions about Queen Elizabeth I. And the readers — oh, the readers. I wanted to get the research straight and the tone right, and doing both meant reaching out to plenty of readers with different passions and areas of expertise. I'm grateful to Jamie Adaway, Joyce Allen, Sven Arvidson, Marlene Berhmann, Victoria Brescoll, Mark Cohan, Alice Eagly, Susan Fisk, Katie Foster, Vicki Helgeson, Richard Hoffman, Jacquelyn Miller, Karen Gee, David Green, Rebecca Jaynes, Susie J. Lee, Junlei Li, Mara Mather, Norma Ming, Julie Nelson, Bryan Ruppert, Michelle Ryan, David Silverman, Giannina Silverman, Ruud van den Bos, Jonathan Weaver, Elanor Williams, and Anita Woolley. To everyone who offered feedback: what you read in this book may bear little resemblance to what you read once upon a time, but that just means I took your constructive comments to heart.

Before this book had a team, it had a club. Tina Zamora and Matt Whitlock have been members of a small but mighty writing club, and since 2010 the three of us have met once a month at a café to discuss our respective writing projects. We talk, we write, and we know the others care.

Several people kept me healthy and productive as I hammered away at my keyboard. Carla Bradshaw, Jennifer Kosaka, Frank Marinkovich,

Jessie Marrs, Robert Martinez, Randip Singh, and Andrei Mousasti-coshvily knew what a toll this book could have taken, and they kept me out of harm's way.

We don't get to pick our families, but I got lucky. The women in my family have made unpopular, tough decisions when the easy options weren't working. My grandmother asked my grandfather for months if they could widen the doorway between the living room and dining room, to no avail, so one day she picked up a sledgehammer and widened that doorway herself. After years of trying to make a marriage work, my sister decided she needed to get a divorce, and as a single working mom, she has raised two daughters to be hard-working, brave, playful, and happy. And then there's my mom. When I called my mother from my dorm room twenty-five years ago, crying and saying I wanted to drop out of college, she didn't try to talk me out of it. She said I could do that if that was what I really wanted, but she would come visit me first, and she was going to buy a plane ticket as soon as we got off the phone. Long pause. Having my mother on campus was worse than anything I was going through, so I said I'd stay and keep trying. She told me, years later, that she was terrified on the phone with me that night, and she'd prayed that leaving the decision up to me was the right thing to do. It was, Mom.

I wouldn't have attempted this book if it weren't for my husband, Jonathan. He agreed to cover our financial needs for several years so I could try my luck as a full-time writer. If my husband's primary gift was the tangible support of time and money, that would have been more than enough. But he also offered me, and continues to offer me, the intangible. He stokes my ambition and reminds me of my own best advice, and he stands a little taller when he tells people about this book. Everyone on this team made me a better writer, but he made me a believer.

Notes

What Happens When a Woman Makes the Call?

1. Hunter Stuart, "Best Buy Ends Work-from-Home Program Known As 'Results Only Work Environment,'" *Huffington Post,* March 6, 2013. The story about Best Buy wasn't completely ignored, but it didn't get the attention that Yahoo's announcement did. *Harvard Business Review* ran a story explaining why Best Buy's decision was a much bigger threat to work-life balance, especially since Joly told investigators that he wanted every employee "to feel dispensable." But the popular press continued to scrutinize Mayer, not Joly; see Monique Valcour, "The End of 'Results Only' at Best Buy Is Bad News," *Harvard Business Review,* March 8, 2013.

2. A quick Internet search will bring up hundreds of articles analyzing whether Mayer made the right decision. Here are two articles that appeared a full two years after she changed Yahoo's work-from-home policy: Nicholas Bloom and John Roberts, "A Working from Home Experiment Shows High Performers Like It Better," *Harvard Business Review,* January 23, 2015, hbr.org/2015/01/a -working-from-home-experiment-shows-high-performers-like-it-better; Akane Otani, "Richard Branson: Marissa Mayer's Yahoo Work Policy Is on the Wrong Side of History," *Bloomberg Business,* April 24, 2015, www.bloomberg .com/news/articles/2015-04-24/richard-branson-marissa-mayer-s-yahoo-work -policy-is-on-the-wrong-side-of-history. Richard Branson, the CEO of Virgin Airlines, was asked whether Mayer made a mistake by eliminating the work-from-home policy; as far as I can tell, he wasn't asked about Hubert Joly's decision.

3. For the number of employees affected by the change in Yahoo's work-from-home policy, see Rebecca Greenfield, "Marissa Mayer's Work-from-Home Ban Is Working for Yahoo, and That's That," *Atlantic*, March 6, 2013. For the number of employees affected by the change in Best Buy's work-from-home policy, see Julianne Pepitone, "Best Buy Ends Work-from-Home Program," CNNMoney.com, March 5, 2013.

4. Mayer became the CEO of Yahoo in the July 2012, and Joly was made the CEO of Best Buy in August of that same summer. It is true that Hubert Joly had been a CEO before he came to Best Buy—he was the CEO of Carlson Wagonlit Travel from 2008 to 2012—whereas Mayer had previously been a vice president at Google. So Joly had more experience as a CEO but not more experience as the CEO of Best Buy.

5. In the late 1960s, not giving women a chance to review their biopsy results was part of the standard treatment of breast cancer. Barron H. Lerner, *The Breast Cancer Wars: Hope, Fear, and the Pursuit of a Cure in Twentieth-Century America* (New York: Oxford University Press, 2001).

6. Barbara Winslow shared the story of her doctor's-office experience in "Primary and Secondary Contradictions in Seattle: 1967–1969," *The Feminist Memoir Project: Voices from Women's Liberation,* eds. Rachel Blau DuPlessis and Ann Snitow (New York: Three Rivers Press, 1998), 227–29. I also spoke with Barbara over the phone on February 4, 2015, and she shared additional details and reflections. Some people might wonder if Barbara just had a terrible doctor, but he was supposed to be excellent and came highly recommended from family members in the area. Her husband was mortified at the thought of signing her medical forms, but, like Barbara, he didn't know what else to do.

7. Richard M. Hoffman et al., "Lack of Shared Decision Making in Cancer Screening Discussions: Results from a National Survey," *American Journal of Preventive Medicine* 47, no. 3 (2014): 251–59. I spoke with Rich Hoffman, the lead investigator, to learn more about his research. Hoffman and his team used survey data collected by another group, the Knowledge Networks. Unfortunately, the data did not include information about the gender of the doctor, so he couldn't analyze whether that made a difference. The purpose of their study was to understand the decision-making process and determine when shared decision-making occurred for cancer screening. The gender of the patient was just one of the variables they analyzed, but it stood out as a significant one.

8. U.S. Preventive Services Task Force, Recommendation Summary, May 2012. The B rating for mammograms is for women who are ages fifty to seventy-four

and get tested once every two years. The D rating for the prostate blood test, called the PSA, for "prostate-specific antigen," appears to be for all age groups.

9. In general, if a PSA value is higher than 5 ng/mL (depending on a man's age), it's considered elevated. However, three out of four men with elevated PSAs don't have prostate cancer, which means the test makes a lot of men and their families worried for nothing; see "Problems with the PSA," Prostate Centre, accessed March 23, 2015, www.theprostatecentre.com/prostate-information /the-psa-test/problems-with-the-psa/. See also "Prostate Cancer Screening: Should You Get the PSA Test?," Mayo Clinic, www.mayoclinic.org/diseases -conditions/prostate-cancer/in-depth/prostate-cancer/art-20048087. There are so many problems with the PSA test that the 2008 U.S. Preventive Services Task Force recommended *against* using it; see Virginia Moyer, "Screening for Prostate Cancer: U.S. Preventive Services Task Force Recommendation Statement," *Annals of Internal Medicine* 157, no. 2 (2012): 120–34.

10. "Colorectal Cancer Screening: Recommendation Summary," U.S. Preventive Services Task Force (October 2008). The A rating for colonoscopies applies to adults ages fifty to seventy-five.

11. Rebecca L. Siegel, Kimberly D. Miller, and Ahmedin Jemal, "Cancer Statistics, 2015," *CA: A Cancer Journal for Clinicians* 65, no. 1 (2015): 5–29. See table 1 for estimated deaths by sex. Lung and bronchial cancer is the number-one cause of cancer deaths for both men and women; prostate cancer is number two for men and breast cancer is number two for women.

12. Ibid. See table 4 for men's and women's probability of developing colon cancer.

13. For information on dates that other countries granted women the right to vote, see "Women's Suffrage: A World Chronology of the Recognition of Women's Rights to Vote and to Stand for Election," Inter-Parliamentary Union, accessed June 1, 2015, www.ipu.org/wmn-e/suffrage.htm.

14. It wasn't until 1974 that Congress finally banned discrimination in lending based on sex and marital status. The bill was called the Equal Credit Opportunity Act (ECOA) and President Ford signed it into law on October 28, 1974. Women were now protected, but calling their rights "equal" was a bit of a stretch. Lending institutions could still discriminate based on race, color, ethnicity, religion, national origin, or age. Two years later, in 1976, the ECOA was amended to ensure that lenders couldn't discriminate on these grounds either. Stories about women, including Billie Jean King, being denied mortgages and credit cards are presented in Gail Collins's captivating book *When Everything Changed: The Amazing Journey of American Women from 1960 to the Present* (Boston: Little, Brown, 2009).

15. Selena Roberts, *A Necessary Spectacle* (New York: Crown, 2005), 66. The Equal

Credit Opportunity Act meant that if a woman had the ability to pay, the banks had to trust her as much as they trusted a man.

16. Victoria Brescoll and Eric Luis Uhlmann, "Can an Angry Woman Get Ahead? Status Conferral, Gender, and Expression of Emotion in the Workplace," *Psychological Science* 19, no. 3 (2008): 268–75.

17. In study 2, Brescoll and Uhlmann specifically asked participants to rate, on a scale of 1 to 11, whether the job candidate was an "in control" person (a score of 1) or "out of control" (a score of 11). The angry female candidates were seen as significantly more out of control than the angry male candidates. For instance, the high-ranking, highly experienced female job candidate who got angry received an average score of 6.41 on the out-of-control scale, while the high-ranking, highly experienced male job candidate received an average score of 4.12. (See pages 271–72 in the cited article for the detailed analyses.)

18. It's worth noting here, though, that the job candidates in Brescoll's study were all white; subsequent research suggests that a candidate's racial background also figures strongly into how assertiveness, anger, or aggression influences ranking. To the best of my knowledge, no one has replicated Brescoll's study using black job applicants, but Robert W. Livingston of the Kellogg School of Management at Northwestern University has done a number of interesting studies showing that race and gender shape how people determine whether a person should lead and be given more power or have power taken away. In one study, Livingston and his colleagues found that people penalized white female and black male leaders who expressed dominance and assertiveness. They thought these leaders should be paid less and believed they were less effective at their jobs than white females and black male leaders who had more encouraging and caring leadership styles. White male leaders and, interestingly enough, black female leaders weren't penalized when they behaved in dominant, aggressive ways. See Robert W. Livingston, Ashleigh Shelby Rosette, and Ella F. Washington, "Can an Agentic Black Woman Get Ahead? The Impact of Race and Interpersonal Dominance on Perceptions of Female Leaders," *Psychological Science* 23, no. 4 (2012): 354–58. Livingston also coined the term the *teddy-bear effect*; he and Nicholas Pearce found that black male CEOs, though few and far between, were rated as being more "baby-faced" than white CEOs. Having a baby face has been shown to be a problem for white men who want to rise through the ranks, but it appears to help black men. Why? The current interpretation is that having a baby face makes a black man appear warmer and less threatening, which helps others accept him as an authority figure; see Robert W. Livingston and Nicholas A. Pearce, "The Teddy-Bear Effect: Does Having a Baby Face Benefit Black Chief Executive Officers?," *Psychological Science* 20, no. 10 (2009): 1229–33.

19. Only 15 percent of the executives in Fortune 500 companies, the five hundred largest companies on the New York Stock Exchange, are women. See Catalyst's Quick Take, www.catalyst.org/knowledge/women-united-states. Likewise, 15 percent of the executives in the FTSE 100, the one hundred largest companies on the London Stock Exchange, are women. See Ruth Sealy and Susan Vinnicombe, *The Female FTSE Board Report* (Cranfield, UK: Cranfield University, 2013), 17.

20. Justin Wolfers, "Fewer Women Run Big Companies Than Men Named John," *New York Times*, March 2, 2015, www.nytimes.com/2015/03/03/upshot/fewer -women-run-big-companies-than-men-named-john.html?_r=0&abt =0002&abg=1.

21. Of the midlevel and senior managers who responded to the question about wanting a top management position in the C-suite, 81 percent of the men and 79 percent of women said they wanted that job. These data are reported in Sandrine Devillard et al., *Women Matter 2013: Gender Diversity in Top Management* (Paris: McKinsey and Company, 2013), 10. When senior female managers were asked if they wanted to move up in their organization, 51 percent said they had a strong desire to do so, and 32 percent said they had some desire, for a total of 83 percent of women who wanted to move to the next level. When senior male managers were asked the same question, 37 percent said they had a strong desire and 37 percent said they had some desire, for a total of 74 percent of men who wanted to move up to the next level. See Sandrine Devillard and Sandra Sancier-Sultan, *Moving Mindsets on Gender Diversity* (Paris: McKinsey and Company, 2013), 2.

22. For the study of 3,345 MBAs, see Nancy M. Carter and Christine Silva, "The Myth of the Ideal Worker: Does Doing All the Right Things Really Get Women Ahead?," Catalyst, 2011, http://www.catalyst.org/knowledge/myth-ideal-worker-does-doing-all-right-things-really-get-women-ahead. For data on how networking benefits men on Wall Street but does little for women, see Lily Fang and Sterling Huang, "Gender and Connections Among Wall Street Analysts," working paper, February 27, 2015, www.insead.edu/facultyresearch /research/doc.cfm?did=48816.

23. In some fields, the small number of women at the top can be explained by the fact that there have historically been fewer women even at the lower levels. Take chess, instance: As of 2004, only 1 percent of chess grand masters were women. Bias in the system? Not likely. Chess ratings are based on objective scoring, so there is no "old boys' network" to keep women from rising through the ranks. Christopher Chabris and Mark Glickman studied a database of over 256,000 tournament chess players and found that "the greater number of men

at the highest levels of chess can be explained by the greater number of boys who enter chess at the lowest levels"; see Christopher F. Chabris and Mark E. Glickman, "Sex Differences in Intellectual Performance," *Psychological Science* 17, no. 12 (2006): 1040. So when people are ranked purely on merit and few young women pursue a field, a slow pipeline can explain the difference at the top. But as we'll see later in the book, men and women can do the same thing in the workplace and be ranked very differently.

24. From 1995 to 2005, there was promising growth in the number of Fortune 500 board seats held by women, from 9.6 percent to 14.7 percent, growing at a rate of half a percentage point a year. And then progress stalled. Over the next nine years, from 2005 to 2014, the percentage of women board members crawled from 14.7 percent to 16.9 percent, less than half of the growth of the decade before. For data from 1995 to 2013, see "Statistical Overview of Women in the Workplace," Catalyst (March 3, 2014), www.catalyst.org/knowledge/statistical -overview-women-workplace. See especially the figure labeled "Fortune 500 Seats Held by Women." For 2014 data, see Caroline Fairchild, "A Call to Action for Companies with No Female Directors," Fortune.com, November 14, 2014. In 2015, Catalyst stopped tracking the Fortune 500 and began reporting S&P 500 data. The number of female board members on the S&P 500 was 19.2 percent as of 2014, but because it's a different group of companies, it's not possible to make a direct comparison.

25. For data on the number of state governors, see Pew Research Center, "Women and Leadership: Public Says Women Are Equally Qualified, but Barriers Persist" (January 2015), http://www.pewsocialtrends.org/2015/01/14/women -and-leadership/12.

26. Sylvia Ann Hewlett et al., *The Athena Factor: Reversing the Brain Drain in Science, Engineering, and Technology* (Watertown, MA: Harvard Business School, 2008).

27. Kieran Snyder, "Why Women Leave Tech: It's the Culture, Not Because 'Math Is Hard,'" Fortune.com, October 2, 2014, fortune.com/2014/10/02/women -leave-tech-culture/. Kieran Snyder provided this quote in an e-mail communication with the author on March 24, 2015.

28. Pew Research Center, "Women and Leadership," 17.

29. John Gerzema and Michael D'Antonio, *The Athena Doctrine* (San Francisco: Jossey-Bass, 2013).

30. Bureau of Labor Statistics, "Table 11: Employed Persons by Detailed Occupation, Sex, Race, and Hispanic or Latino Ethnicity," *Current Population Survey, Annual Averages 2014* (Washington, DC: United States Department of Labor, 2014). In 2014, women made up 51.6 percent of "management,

professional, and related occupations" in the United States. This statistic is often cited, but that "professional" umbrella includes teachers and nurses, which inflates the percentage. When we look at management occupations alone, the number drops to 38.6 percent.

31. "Women CEOs of the S&P 500," Catalyst, 2015, http://www.catalyst.org /knowledge/women-ceos-sp-500.

32. Seth J. Prins et al., "Anxious? Depressed? You Might Be Suffering from Capitalism: Contradictory Class Locations and the Prevalence of Depression and Anxiety in the USA," *Sociology of Health and Illness* 37, no. 8 (November 2015): 1352–72.

33. For a detailed and fascinating history of men and women in computer programming, see Isaacson's *The Innovators* as well as Laura Sydell, "The Forgotten Female Programmers Who Created Modern Tech," *All Things Considered,* NPR, October 6, 2014, www.npr.org/sections/alltechconsidered /2014/10/06/345799830/the-forgotten-female-programmers-who-created -modern-tech. Jean Jennings Bartik, one of the six women programmers on the first general-purpose computer, said that if the male "administrators had known how crucial programming would be to the functioning of the electronic computer and how complex it would be, they might have been more hesitant to give such an important role to women." Jean Jennings Bartik, *Pioneer Programmer* (Kirksville, MO: Truman State University Press, 2013), 557.

34. Steve Henn, "When Women Stopped Coding," *Planet Money,* NPR, October 21, 2014, www.npr.org/sections/money/2014/10/21/357629765/when-women -stopped-coding.

35. This list of decision-making skills is adapted from Chip and Dan Heath's fantastic book *Decisive: How to Make Better Decisions in Life and Work* (New York: Crown Business, 2013).

36. Joan C. Williams and Rachel Dempsey, *What Works for Women at Work* (New York: New York University Press, 2014).

37. This point was beautifully argued by Sarah Green Carmichael in her article "Why 'Network More' Is Bad Advice for Women," *Harvard Business Review,* February 26, 2015, hbr.org/2015/02/why-network-more-is-bad-advice-for -women.

1. Making Sense of Women's Intuition

1. Audrey Nelson, "What's Behind Women's Intuition?," *Psychology Today,* February 22, 2015, www.psychologytoday.com/blog/he-speaks-she-speaks /201502/what-s-behind-women-s-intuition.

2. Most of the questions that I've used have been adapted from the Cognitive

Style Index, a psychometric test that many academic researchers use to determine whether a person takes an intuitive or analytical approach to gathering, processing, and interpreting information. For the original paper explaining the Cognitive Style Index, see Christopher W. Allinson and John Hayes, "The Cognitive Style Index: A Measure of Intuition-Analysis for Organizational Research," *Journal of Management Studies* 33, no. 1 (1996): 119–35. For more information and sample questions from the Cognitive Style Index, see Christopher W. Allinson and John Hayes, *The Cognitive Style Index: Technical Manual and User Guide* (Harlow, UK: Pearson Education, 2012), and "Gender and Learning," AONTAS, 2003, www.aontas.com /pubsandlinks/publications/gender-and-learning-2003/. I have also adapted a few questions from the intuitive-sensing scale used in the Myers-Briggs test described by Charles R. Martin, *Looking at Type: The Fundamentals* (Gainesville, FL: Center for Applications of Psychological Type, 2001). Some readers might be wondering how this questionnaire is different from the Myers-Briggs. Indeed, the Myers-Briggs is a popular personality test and it's often used in corporate settings to measure how people perceive and experience their work and relationships. The Myers-Briggs tests four dimensions, one of which is the intuitive-sensing dimension. If you pay more attention to the patterns, theories, and possibilities in your own mind than to those in the outside world, the Myers-Briggs would say you have an intuitive style (the *N* in Myers-Briggs coding); if you prefer to focus on information coming in from the world around you, the Myers-Briggs would say you have a sensing style (the *S* in Myers-Briggs coding). It is possible that you would score as analytical on my questionnaire and intuitive on the Myers-Briggs. Keep in mind that mine is just an informal set of questions. But the other important difference is what's meant by *intuitive*. In my questionnaire, which draws heavily on the Cognitive Style Index, *intuitive* means being relatively spontaneous and basing one's impression on wholes, while *analytical* means being linear, sequential, and focused on detail. That's rather different from what *intuitive* means on the Myers-Briggs. On the Myers-Briggs, *intuitive* means you trust your own thoughts more than external information and you prefer to learn by thinking. You might be an extremely linear thinker, one who is focused on the details, which means you'd be analytical according to my questionnaire, but if you'd rather learn by thinking than by doing, it would make you intuitive on the Myers-Briggs. Research psychologists are rather mixed on the Myers-Briggs tests. Industrial and organizational psychologists tend to have a warmer view of the Myers-Briggs, but personality psychologists often find it deeply problematic and unreliable. For one analysis of the Myers-Briggs test as a measure of personality, see Robert R. McCrae and Paul T. Costa, "Reinterpreting the Myers-Briggs Type Indicator from the Perspective of the Five-Factor Model of Personality," *Journal of Personality* 57, no. 1 (1989): 17–40.

3. Quoted in Sonia Choquette, *The Time Has Come . . . to Accept Your Intuitive Gifts* (London: Hay House, 2008), 93.

4. Oprah Winfrey, "What Oprah Knows for Sure About Trusting Her Intuition," *O, the Oprah Magazine*, August 2011, www.oprah.com/spirit/Oprah-on -Trusting-Her-Intuition-Oprahs-Advice-on-Trusting-Your-Gut.

5. Quoted in Harvey A. Dorfman, *Coaching the Mental Game* (Lanham, MD: Taylor Trade, 2003), 146.

6. Martin Robson, "Feeling Our Way with Intuition," in *Bursting the Big Data Bubble: The Case for Intuition-Based Decision Making*, ed. Jay Liebowitz (Boca Raton, FL: Taylor and Francis, 2014), 23.

7. Erik Dane, Kevin W. Rockmann, and Michael G. Pratt, "When Should I Trust My Gut? Linking Domain Expertise to Intuitive Decision-Making Effectiveness," *Organizational Behavior and Human Decision Processes* 119, no. 2 (2012): 187–94.

8. Gary Klein, *Seeing What Others Don't: The Remarkable Ways We Gain Insights* (New York: Public Affairs, 2013), 26.

9. The story was originally told by Gary Klein but captured beautifully in the work of Daniel Kahneman, *Thinking, Fast and Slow* (New York: Farrar, Straus and Giroux, 2011), 11–12.

10. For the four-part definition of intuition, see Erik Dane and Michael G. Pratt, "Exploring Intuition and Its Role in Managerial Decision-Making," *Academy of Management Review* 32, no. 1 (2007): 33–54, and Erik Dane and Michael G. Pratt, "Conceptualizing and Measuring Intuition: A Review of Recent Trends," *International Review of Industrial and Organizational Psychology* 24 (2009): 1–40.

11. Allinson and Hayes, *The Cognitive Style Index*. However, at least one study using the Myers-Briggs Type Indicator's Intuitive scale reports that women managers score higher on that scale than male managers; see Weston H. Agor, *The Logic of Intuitive Decision-Making: A Research-Based Approach for Top Management* (New York: Quorum Books, 1986). Details on the differences between Myers-Briggs and the Cognitive Style Index are offered in note 2 for this chapter.

12. Cecilia L. Ridgeway and Lynn Smith-Lovin, "The Gender System and Interaction," *Annual Review of Sociology* (1999): 191–216.

13. For a general description of this research, see Christopher F. Karpowitz and

Tali Mendelberg, *The Silent Sex: Gender, Deliberation, and Institutions* (Princeton, NJ: Princeton University Press, 2014), 72–73. For the original research, see Meredith D. Pugh and Ralph Wahrman, "Neutralizing Sexism in Mixed-Sex Groups: Do Women Have to Be Better Than Men?," *American Journal of Sociology* (1983): 746–62. See also Cecilia L. Ridgeway, "Status in Groups: The Importance of Motivation," *American Sociological Review* (1982): 76–88.

14. John, Hayes, Christopher W. Allinson, and Steven J. Armstrong, "Intuition, Women Managers and Gendered Stereotypes," *Personnel Review* 33, no. 4 (2004): 403–17. See also W. M. Taggart et al., "Rational and Intuitive Styles: Commensurability Across Respondents' Characteristics," *Psychological Reports* 80, no. 1 (1997): 23–33.

15. John Coates, *The Hour Between Dog and Wolf* (New York: Penguin, 2012).

16. My definition of *interpersonal sensitivity* comes from Sara D. Hodges, Sean M. Laurent, and Karyn L. Lewis, "Specially Motivated, Feminine, or Just Female: Do Women Have an Empathic Accuracy Advantage?," in *Managing Interpersonal Sensitivity: Knowing When — and When Not — to Understand Others*, eds. J. L. Smith et al. (New York: Nova Science Publishers, 2011), 59–74.

17. Cordelia Fine, *Delusions of Gender: How Our Minds, Society, and Neurosexism Create Difference* (New York: W. W. Norton, 2010), 17–18. See also Carol Tavris, *The Mismeasure of Women* (New York: Simon and Schuster, 1992).

18. Fine, *Delusions of Gender*.

19. Louann Brizendine, *The Female Brain* (London: Bantam, 2007), 161.

20. Judith A. Hall, "Gender Effects in Decoding Nonverbal Cues," *Psychological Bulletin* 85, no. 4) (1978): 845–57; Judith A. Hall, *Nonverbal Sex Differences: Communication Accuracy and Expressive Style* (Baltimore: Johns Hopkins University Press, 1984).

21. The Reading the Mind in the Eyes Test was developed and popularized by a team of University of Cambridge professors led by Simon Baron-Cohen. Baron-Cohen and his colleagues were looking for a test that correlated strongly with autistic symptoms and found that autistic adults performed quite poorly on the Eyes Test. But Anita Williams Woolley and her colleagues were the first to use the test to assess team performance. For the original study by the Cambridge team, see Simon Baron-Cohen et al., "The 'Reading the Mind in the Eyes' Test, Revised Version: A Study with Normal Adults, and Adults with Asperger Syndrome or High-Functioning Autism," *Journal of Child Psychology and Psychiatry* 42, no. 2 (2001): 241–51.

22. Anita Williams Woolley et al., "Evidence for a Collective Intelligence Factor in

the Performance of Human Groups," *Science* 330, no. 6004 (2010): 686–88. See also David Engel et al., "Reading the Mind in the Eyes or Reading Between the Lines? Theory of Mind Predicts Collective Intelligence Equally Well Online and Face-to-Face," *PLOS ONE* 9, no. 12 (2014), e115212.

23. Kristi J. K. Klein and Sara D. Hodges, "Gender Differences, Motivation, and Empathic Accuracy: When It Pays to Understand," *Personality and Social Psychology Bulletin* 27, no. 6 (2001): 720–30. Klein and Hodges asked men and women to watch a video and then complete a questionnaire identifying the emotions the person in the video was feeling. Women showed much higher interpersonal sensitivity than men when they thought empathy was being measured, but when the subjects thought cognitive ability was being measured, men and women scored equally well. To the best of my knowledge, no one has tried to motivate men to improve their performance on the Eyes Test.

24. Geoff Thomas and Gregory R. Maio, "Man, I Feel Like a Woman: When and How Gender-Role Motivation Helps Mind-Reading," *Journal of Personality and Social Psychology* 95, no. 5 (2008): 1165.

25. Some researchers argue that men can be motivated to perform as well as women but men's average ability to read others doesn't exceed women's, even when men are highly motivated. See Judith A. Hall and Marianne Schmid Mast, "Are Women Always More Interpersonally Sensitive Than Men? Impact of Goals and Content Domain," *Personality and Social Psychology Bulletin* 34, no. 1 (2008): 144–55.

26. Hodges, Laurent, and Lewis, "Specially Motivated."

27. For research looking at the relationship between empathy and testosterone levels, see Jack Van Honk et al., "Testosterone Administration Impairs Cognitive Empathy in Women Depending on Second-to-Fourth Digit Ratio," *Proceedings of the National Academy of Sciences* 108, no. 8 (2011): 3448–52. For specific research looking at performance on the Reading the Mind in the Eyes Test related to testosterone levels, see Emma Chapman et al., "Fetal Testosterone and Empathy: Evidence from the Empathy Quotient (EQ) and the 'Reading the Mind in the Eyes' Test," *Social Neuroscience* 1, no. 2 (2006): 135–48.

28. Tavris, *The Mismeasure of Women*, 65.

29. Sara E. Snodgrass, "Women's Intuition: The Effect of Subordinate Role on Interpersonal Sensitivity," *Journal of Personality and Social Psychology* 49, no. 1 (1985): 146–55.

30. Since Snodgrass's original study, the power of subordinates' intuition has been replicated by other researchers; see A. Galinsky et al., "Power and Perspectives

Not Taken," *Psychological Science* 17, no. 12 (2006): 1068–74, and David Kenny et al., "Interpersonal Sensitivity, Status, and Stereotype Accuracy," *Psychological Science* 21, no. 12 (2010): 1735–39. For an alternative view, see Dario Bombari et al., "How Interpersonal Power Affects Empathic Accuracy: Differential Roles of Mentalizing vs. Mirroring," *Frontiers in Human Neuroscience* 7 (2013).

31. For an accessible article describing Woolley's line of research on collective intelligence, see Anita Woolley, Thomas Malone, and Christopher Chabris, "Why Some Teams Are Smarter Than Others," *New York Times,* January 8, 2015. For the original research studies, see Woolley et al., "Evidence for a Collective Intelligence Factor." See also Engel et al., "Reading the Mind in the Eyes."

32. For the details of Woolley's methods, see the supplemental online materials that accompany her 2010 study. For work on how women's contributions are marginalized, see Karpowitz and Mendelberg, *The Silent Sex,* 143. For a skeptical analysis of the impact of team diversity, see Alice H. Eagly, "When Passionate Advocates Meet Research on Diversity: Does the Honest Broker Stand a Chance?," *Journal of Social Issues* (in press).

33. Dan Ariely describes this study charmingly in his book *Predictably Irrational* (New York: HarperCollins, 2008), 26–29. The original research study is Dan Ariely, George Loewenstein, and Drazen Prelec, "'Coherent Arbitrariness': Stable Demand Curves Without Stable Preferences," *Quarterly Journal of Economics* 118, no. 1 (2003): 73–105.

34. For research on how hard it can be to avoid the anchoring effect, see Joseph P. Simmons, Robyn A. LeBoeuf, and Leif D. Nelson, "The Effect of Accuracy Motivation on Anchoring and Adjustment: Do People Adjust from Provided Anchors?," *Journal of Personality and Social Psychology* 99, no. 6 (2010): 917–32.

35. R. M. Hogarth, *Educating Intuition* (Chicago: University of Chicago Press, 2001).

36. This definition of a kind environment is offered by Chip Heath and Dan Heath in *Decisive: How to Make Better Decisions in Life and Work* (New York: Crown Business, 2013), 277.

37. Thankfully, technology such as clickers and classroom response systems are making it possible to receive immediate feedback on what students do and don't understand in class. You can teach something and then ask a question five minutes later to test students' comprehension. That greatly improves the speed of feedback, but it doesn't mean that you get clear feedback on what part of your teaching worked and what part didn't.

38. James Shanteau, "Competence in Experts: The Role of Task Characteristics," *Organizational Behavior and Human Decision Processes* 53, no. 2 (1992): 252–66. For an interesting book on how stockbrokers can develop skilled intuitions, see Coates, *The Hour Between Dog and Wolf.*

39. This contrast between kind and wicked medical environments is taken from Daniel Kahneman and Gary Klein's article "Conditions for Intuitive Expertise: A Failure to Disagree," *American Psychologist* 64, no. 6 (2009): 522.

40. Gary Klein interview, "Strategic Decisions: When Can You Trust Your Gut?," *McKinsey Quarterly* (March 2010): 58–67.

41. "Ice Cream Sales and Trends," International Dairy Foods Association, accessed August 7, 2015, www.idfa.org/news-views/media-kits/ice-cream/ice-cream-sales-trends.

42. Adam L. Alter et al., "Overcoming Intuition: Metacognitive Difficulty Activates Analytic Reasoning," *Journal of Experimental Psychology: General* 136, no. 4 (2007): 569.

43. If you'd like to learn more about different ways to trigger intuitive and analytical modes of decision-making, Nina Horstmann, Daniel Hausmann, and Stefan Ryf published a nice summary of different methods in "Methods for Inducing Intuitive and Deliberate Processing Modes," in *Foundations for Tracing Intuition: Challenges and Methods,* eds. Andreas Glöckner and Cilia Witteman (New York: Psychology Press, 2009), 219–37.

44. Adam K. Fetterman and Michael D. Robinson, "Do You Use Your Head or Follow Your Heart? Self-Location Predicts Personality, Emotion, Decision Making, and Performance," *Journal of Personality and Social Psychology* 105, no. 2 (2013): 316. It's interesting to note that Fetterman and Robinson did find that 58 to 66 percent of women said that they prefer to "go with their heart" while 54 to 78 percent of men said they prefer to "go with their head" when it came to making choices. The exact percentages varied from experiment to experiment, but the basic trend was always the same, more men than women going with their heads. So whether or not one finds a gender difference in decision-making seems to lie in how one asks the question. If the question is "Do you go with your heart or your head?," then you'll see a big gender divide that's consistent with the stereotype that women prefer an intuitive approach and men prefer a more analytical approach. But if we ask the kinds of questions on pages 32–33, questions about more specific behaviors, such as "How frequently do you make off-the-top-of-the-head decisions?" and "How often do you scan through reports rather than reading them in detail?," then

we don't find men and women lining up with stereotypical decision-making strategies. Perhaps what's stereotypical isn't so much how one makes decisions but how people like to think of themselves.

45. For one oft-cited paper on the flaws in embodied cognition, see Bradford Z. Mahon and Alfonso Caramazza, "A Critical Look at the Embodied Cognition Hypothesis and a New Proposal for Grounding Conceptual Content," *Journal of Physiology — Paris* 102, no. 1 (2008): 59–70.

46. A look-back is very similar to a strategy that is discussed in chapter 3, the premortem, which was developed by Daniel Kahneman; see his *Thinking, Fast and Slow*.

47. Heath and Heath, *Decisive*, 15.

48. Robert L. Dipboye, "Structured and Unstructured Selection Interviews: Beyond the Job-Fit Model," in *Research in Personnel and Human Resources Management*, ed. Gerald Ferris (Greenwich, CT: JAI Press, 1994), 79–123.

49. David G. Myers, *Intuition: Its Powers and Perils* (New Haven, CT: Yale University Press, 2002). See also Heath and Heath, *Decisive*.

50. Jason Dana, Robyn Dawes, and Nathanial Peterson, "Belief in the Unstructured Interview: The Persistence of an Illusion," *Judgment and Decision Making* 8, no. 5 (2013): 512–20.

51. Robyn M. Dawes, *House of Cards: Psychology and Psychotherapy Built on Myth* (New York: Free Press, 1994).

52. Hiring questions adapted from Heath and Heath, *Decisive*.

2. The Decisiveness Dilemma

1. Bryan A. Reaves, "Census of State and Local Law Enforcement Agencies, 2008," U.S. Department of Justice, July 2011, accessed October 23, 2015, http://www.bjs.gov/content/pub/pdf/csllea08.pdf. See table 2 for sizes of various state and local police departments across the United States. A local police department with over five hundred officers is huge and atypical in the United States. Almost half (48 percent) of all local police departments have fewer than ten officers, as of 2013; see Bryan A. Reaves, "Local Police Departments, 2013: Personnel, Policies, and Practices," U.S. Department of Justice, May 2015, accessed June 21, 2015, www.bjs.gov/content/pub/pdf/lpd13ppp.pdf.

2. Lynn Langton, "Women in Law Enforcement: 1987–2008," Bureau of Justice Statistics, June 2010. The 15 percent statistic is taken from figure 2 on page 1. The report looks at the gender balance in a variety of different law enforcement offices, from the Criminal Investigation arm of the Internal Revenue Service (a whopping 32 percent women) all the way down to small, local police departments (which average only 4 percent women). Although it's

not true in all cases, the general pattern seems to be that the smaller the department, the greater the proportion of men.

3. International Association of Chiefs of Police and National Association of Women Law Enforcement Executives, *Women in Law Enforcement Survey* (Lenexa, KS: NAWLEE, 2013).

4. George P. Monger, "Breach of Promise," in *Marriage Customs of the World: An Encyclopedia of Dating Customs and Wedding Traditions*, 2nd ed., vol. 2 (Santa Barbara, CA: ABC-CLIO, 2013), 85–87. For a discussion of the theory that the law was based on assumptions that the woman might have given up her virginity once a man promised marriage, see Margaret F. Brinig, "Rings and Promises," *Journal of Law, Economics, and Organization* 6, no. 1 (Spring 1990): 203–15.

5. For data on the average breach-of-promise settlement in the 1850s, see Denise Bates, *Breach of Promise to Marry: A History of How Jilted Brides Settled Scores* (South Yorkshire, UK: Wharncliffe, 2014), 96. To calculate how much income you'd need in 2014 to have the same purchasing power as £390, I first used the incredibly handy calculator "Purchasing Power of British Pounds from 1270 to Present," maintained by Measuring Worth, accessed August 9, 2015, www .measuringworth.com/ppoweruk/. That gave me the income value of £530,700. I then used the average exchange rate for converting the UK pound to the U.S. dollar for 2014.

6. For an example of the media presenting Barbara Bush's statement as a woman's prerogative, see Elisabeth Parker, "Whoops: Barbara Bush Changes Her Mind About Jeb," AddictingInfo.org, accessed May 25, 2015, www.addictinginfo .org/2015/03/18/barbara-bush-endorses-jeb-bush-2016-run-video/.

7. Kurt Donaldson, "Matthews Asked: Is Hillary Clinton Unable to 'Admit a Mistake' on Iraq Vote Because She Would Be Criticized as a 'Fickle Woman'?," MediaMatters.org, March 17, 2006, mediamatters.org/research/2006/03/17 /matthews-asked-is-hillary-clinton-unable-to-adm/135150.

8. For the 2015 study, see Pew Research Center, "Women and Leadership: Public Says Women Are Equally Qualified, but Barriers Persist" (January 2015), http:// www.pewsocialtrends.org/2015/01/14/women-and-leadership/12.

9. For the "typical American" studies, see Deborah A. Prentice and Erica Carranza, "What Women and Men Should Be, Shouldn't Be, Are Allowed to Be, and Don't Have to Be: The Contents of Prescriptive Gender Stereotypes," *Psychology of Women Quarterly* 26, no. 4 (2002): 269–81. For women, *decisive* was seen as number 33 out of 43 adjectives describing the typical woman.

10. Thomas J. Peters and Robert H. Waterman, *In Search of Excellence: Lessons from America's Best-Run Companies* (New York: Harper Business, 2006).

11. Ibid., 120.

12. For the distinction between indecision and indecisiveness, see Annamaria Di Fabio et al., "Career Indecision Versus Indecisiveness: Associations with Personality Traits and Emotional Intelligence," *Journal of Career Assessment* 21, no. 1 (2013): 42–56.

13. Ibid., 43. For similar definitions, see Noa Saka, Itamar Gati, and Kevin R. Kelly, "Emotional and Personality-Related Aspects of Career-Decision-Making Difficulties," *Journal of Career Assessment* 19, no. 1 (February 2011): 3–20.

14. Mike Allen and David S. Broder, "Bush's Leadership Style: Decisive or Simplistic?," *Washington Post*, August 30, 2004.

15. For the research study on the relationship between perceptions of decisiveness and voting behavior, see Ethlyn A. Williams et al., "Crisis, Charisma, Values, and Voting Behavior in the 2004 Presidential Election," *Leadership Quarterly* 20 (2009): 70–86. For popular accounts of how decisiveness swayed voters, see John Harwood, "Flip-Flops Are Looking Like a Hot Summer Trend," *New York Times*, June 23, 2008, www.nytimes.com/2008/06/23/us/politics/23caucus. html?_.

16. Felix C. Brodbeck et al., "Cultural Variation of Leadership Prototypes Across 22 European Countries," *Journal of Occupational and Organizational Psychology* 73 (2000): 1–29.

17. International Association of Chiefs of Police and National Association of Women Law Enforcement Executives, *Women in Law Enforcement Survey*.

18. Cathy Benko and Bill Pelster, "How Women Decide," *Harvard Business Review* (September 2013): 81.

19. David Bakan was the first to use the terms *agency* and *communion* to reflect two modalities for how humans can exist; see David Bakan, *The Duality of Human Existence: An Essay on Psychology and Religion* (Chicago: Rand McNally, 1966). As gender researcher Vicki Helgeson describes it, "Agency reflects one's existence as an individual, and communion reflects the participation of an individual in a larger organism, of which he or she is a part." Bakan proposed that agency was the principle guiding men, whereas communion guided women. See Vicki S. Helgeson and Dianne K. Palladino, "Implications of Psychosocial Factors for Diabetes Outcomes Among Children with Type 1 Diabetes: A Review," *Social and Personality Psychology Compass* 6, no. 3 (2012): 228–42; quote on page 228. In the years since Bakan's work, psychologists have mapped out the achievement-oriented traits associated with agency, such as competence, aggression, independence, forcefulness, and decisiveness, and the relationship-oriented traits associated with communion, such as kindness, helpfulness, and sympathy for others. For research on such

traits and how they are stereotyped masculine and feminine, respectively, see Andrea E. Abele, "The Dynamics of Masculine-Agentic and Feminine-Communal Traits: Findings from a Prospective Study," *Journal of Personality and Social Psychology* 85, no. 4 (2003): 768–76. See also Susan T. Fiske and Laura E. Stevens, "What's So Special About Sex? Gender Stereotyping and Discrimination," in *Gender Issues in Contemporary Society: Applied Social Psychology Annual,* eds. S. Oskamp and M. Costanzo (Newbury, CA: Sage Publications, 1993), 183–96.

20. Jeanine L. Prime, Nancy M. Carter, and Theresa M. Welbourne, "Women 'Take Care,' Men 'Take Charge': Managers' Stereotypic Perceptions of Women and Men Leaders," *Psychologist-Manager Journal* 12 (2009): 25–49.

21. These are sample questions from the General Decision-Making Style Inventory, a survey that is used by many researchers today to measure one's decision-making style; see Suzanne Scott and Reginald Bruce, "Decision-Making Style: The Development and Assessment of a New Measure," *Educational and Psychological Measurement* 55, no. 5 (1995): 818–31.

22. For studies finding that men and women are equally indecisive, see Robert Loo, "A Psychometric Evaluation of the General Decision-Making Style Inventory," *Personality and Individual Differences* 29 (2000): 895–905; also see the various international studies cited below.

23. For Turkey, see Enver Sari, "The Relations Between Decision-Making in Social Relationships and Decision-Making Styles," *World Applied Sciences Journal* 3, no. 3 (2008): 369–81. For Canada, see Lia M. Daniels et al., "Relieving Career Anxiety and Indecision: The Role of Undergraduate Students' Perceived Control and Faculty Affiliations," *Social Psychology Education* 14 (2011): 409–29. For a comparison of gender differences in China, Japan, and the United States, see J. Frank Yates et al., "Indecisiveness and Culture: Incidence, Values, and Thoroughness," *Journal of Cross-Cultural Psychology* 41, no. 3 (2010): 428–44. For a comparison of China and the United States, see Andrea L. Patalano, and Steven M. Wengrovitz, "Cross-Cultural Exploration of the Indecisiveness Scale: A Comparison of Chinese and American Men and Women," *Personality and Individual Differences* 45 (2006): 813–24. For Australia, see G. Beswick, E. D. Rothblum, and L. Mann, "Psychological Antecedents of Student Procrastination," *Australian Psychologist* 23 (1988): 207–17. For New Zealand, see A. L. Guerra and J. M. Braungart-Rieker, "Predicting Career Indecision in College Students: The Roles of Identity Formation and Parental Relationship Factors," *Career Development Quarterly* 47, no. 3 (1999): 255–66.

24. Yates and his colleagues wanted to understand why Japanese adults might be less decisive than Chinese or American adults, and they asked participants to

think aloud as they worked through their decisions. One sharp distinction was how thorough people were in their thinking. Japanese adults stepped through more issues to make decisions, thinking through an average of 7.5 ideas, compared to 3.3 ideas for each Chinese adult and 4.5 ideas for each American. See Yates, "Indecisiveness and Culture."

25. For work on gender differences in decisiveness among adolescents, see Veerle Germeijs and Karine Verschueren, "Indecisiveness and Big Five Personality Factors: Relationship and Specificity," *Personality and Individual Differences* 50, no. 7 (2011): 1023–28. Not everyone finds that teenage girls are less decisive than teenage boys, as discussed in John W. Lounsbury, Teresa Hutchens, and James M. Loveland, "An Investigation of Big Five Personality Traits and Career Decidedness Among Early and Middle Adolescents," *Journal of Career Assessment* 13, no. 1 (2005): 25–39. One possible explanation is that young boys feel pressure to behave like men, which means they feel pressure to be agentic, decisive, in control, confident, and so on, whereas young girls feel pressure to be communal and responsive to others. So when adolescent boys and girls are asked how firmly they feel about their futures, young boys are inclined to give strong, definitive answers and conform to the masculine stereotype even if they're not sure, while young girls don't have to show that certainty, so young boys may appear more decisive than they really are. For research showing that obsessive-compulsive women tend to be more indecisive than obsessive-compulsive men, see Eric Rassin and Peter Muris, "To Be or Not to Be . . . Indecisive: Gender Differences, Correlations with Obsessive–Compulsive Complaints, and Behavioural Manifestation," *Personality and Individual Differences* 38, no. 5, (2005): 1175–81. In general, neuroticism tends to be highly correlated with indecisiveness, as discussed in Germeijs and Verschueren, "Indecisiveness and Big Five Personality Factors."

26. The exact length of the Appalachian Trail changes each year because of repairs. It was 2,181 miles long in 2011, the year Pharr broke the world record.

27. Sandrine Devillard et al., *Women Matter 2013 — Gender Diversity in Top Management: Moving Corporate Culture, Moving Boundaries* (Paris: McKinsey and Company, 2013), 14.

28. For gender differences in leadership style, see Alice H. Eagly and Blair T. Johnson, "Gender and Leadership Style: A Meta-Analysis," *Psychological Bulletin* 108, no. 2 (1990): 233–56. More recent analyses find the same trend: women tend to be more democratic in their leadership style and more focused on the interpersonal relationship than men. See Marloes L. van Engen and Tineke M. Willemsen, "Sex and Leadership Styles: A Meta-Analysis of Research Published in the 1990s," *Psychological Reports* 94, no. 1 (2004): 3–18.

29. Research on how mayors include citizens and the city council in their budget process is described by Lynne A. Weikart et al., "The Democratic Sex: Gender Differences and the Exercise of Power," *Journal of Women, Politics and Policy* 28, no. 1 (2007): 119–40. For data on time spent in meetings on employee feedback, see Eduardo Melero, "Are Workplaces with Many Women in Management Run Differently?," *Journal of Business Research* 64, no. 4 (2011): 385–93.

30. For data on how women are expected to be more focused on the team than men, see Herminia Ibarra and Otilia Obodaru, "Women and the Vision Thing," *Harvard Business Review* 87, no. 1 (2009): 67. For recent data on how we expect women to be more compassionate, see Pew Research Center, "Women and Leadership," 17.

31. Pew Research Center, "Women and Leadership," 21. Of the 1,835 people surveyed, 34 percent thought that women in high political offices were better at working out compromises than men, while only 9 percent thought men were more skilled in this area. An additional 55 percent thought there was no difference between men and women, and 2 percent gave no answer.

32. For an overview of prejudice against female leaders and how women leaders are penalized when they go against what's expected of feminine women, see Alice H. Eagly and Steven J. Karau, "Role Congruity Theory of Prejudice Toward Female Leaders," *Psychological Review* 109, no. 3 (2002): 573–98. For other studies showing negative assessments autocratic women receive, see Madeline E. Heilman, Caryn J. Block, and Richard F. Martell, "Sex Stereotypes: Do They Influence Perceptions of Managers?," *Journal of Social Behavior and Personality* 10 (1995): 237–52. See also Karen Korabik, Galen L. Baril, and Carol Watson, "Managers' Conflict Management Style and Leadership Effectiveness: The Moderating Effects of Gender," *Sex Roles* 29, nos. 5–6 (1993): 405–20.

33. Edward. S. Lopez and Nurcan Ensari, "The Effects of Leadership Style, Organizational Outcome, and Gender on Attributional Bias Toward Leaders," *Journal of Leadership Studies* 8, no. 2 (2014): 19–37.

34. Eagly and Karau, "Role Congruity Theory."

35. Anit Somech, "The Effects of Leadership Style and Team Process on Performance and Innovation in Functionally Heterogeneous Teams," *Journal of Management* 32, no. 1 (2006): 132–57.

36. James R. Larson, Pennie G. Foster-Fishman, and Timothy M. Franz, "Leadership Style and the Discussion of Shared and Unshared Information in Decision-Making Groups," *Personality and Social Psychology Bulletin* 24, no. 5 (1998): 482–95.

37. Research does suggest that women seek more advice than men. See Michael E.

Addis and James R. Mahalik, "Men, Masculinity, and the Contexts of Help Seeking," *American Psychologist* 58, no. 1 (2003): 5–14. Much of the research on gender differences in help-seeking behavior focuses on how women seek more help and advice on health issues, but there is some research on asking for driving directions. One study found that 26 percent of male drivers wait at least half an hour before asking for help, but the researchers didn't report the percentage of female drivers who wait that long. See Scott Mayerowitz, "Male Drivers Lost Longer Than Women," ABC News, October 26, 2010, abcnews .go.com/.

38. The research team looked at over 2,500 mergers and acquisitions from 1997 to 2010; see Maurice Levi, Kai Li, and Feng Zhang, "Are Women More Likely to Seek Advice Than Men? Evidence from the Boardroom," *Journal of Risk and Financial Management* 8, no. 1 (2015): 127–49.

39. Francesca Gino, "Do We Listen to Advice Just Because We Paid for It? The Impact of Advice Cost on Its Use," *Organizational Behavior and Human Decision Processes* 107, no. 2 (2008): 234–45.

40. Andrew Prahl et al., "Review of Experimental Studies in Social Psychology of Small Groups When an Optimal Choice Exists and Application to Operating Room Management Decision-Making," *Anesthesia and Analgesia* 117, no. 5 (2013): 1221–29.

41. Julie Hirschfield Davis and Matt Apuzzo, "Loretta Lynch, Federal Prosecutor, Will Be Nominated for Attorney General," *New York Times*, November 8, 2014.

42. Nick Tasler, "Just Make a Decision Already," *Harvard Business Review*, October 4, 2013, hbr.org/2013/10/just-make-a-decision-already/.

43. Rachel Croson and Uri Gneezy, "Gender Differences in Preferences," *Journal of Economic Literature* (2009): 448–74.

44. Angela Rollins, "Consultant Gives Tips for Interpreting Gender-Linked Communications Styles," *Catalyst* 16, no. 4 (April 2011), www.isba.org /committees/women/newsletter/2011/04/consultantgivestipsforinterpreting genderlinkedcommunic.

45. Diane M. Bergeron, Caryn J. Bloc, and B. Alan Echtenkamp, "Disabling the Able: Stereotype Threat and Women's Work Performance," *Human Performance* 19, no. 2 (2006): 133–58.

46. I found it fascinating, and perhaps a bit ironic, that in a study asking where men and women differ in their approaches to decision-making, the assumption was that it was a high compliment to call women intuitive and men decisive. The researchers had previously tested these adjectives so they knew that each word fit the mold for female and male leaders, respectively.

This simply reaffirms that regardless of how leaders truly behave, people expect female and male leaders to reach decisions differently.

47. Michael Inzlicht and Toni Schmader, eds., *Stereotype Threat: Theory, Process and Application* (New York: Oxford University Press, 2012), 3–14.

48. Stereotype threat is more formally defined as the discomfort and performance anxiety a group member feels "when they are at risk of fulfilling a negative stereotype about their group." Joshua Aronson, Diane M. Quinn, and Steven J. Spencer, "Stereotype Threat and the Academic Underperformance of Minorities and Women," in *Prejudice: The Target's Perspective,* eds. J. K. Swim and C. Strangor (San Diego: Academic Press, 1998), 83–103; quote is on page 85. The poignant "fear of failing" language was popularized by Ed Yong in "Armor Against Prejudice," *Scientific American* (2013).

49. Data from a study released by the Insurance Institute for Highway Safety, www .statisticbrain.com/male-and-female-driving-statistics/, accessed January 15, 2013. The fatal-car-crash rate is nearly the same for men and women once they're seventy years old, but for drivers ages sixteen to sixty-five, men are the more dangerous ones.

50. For the classic study on stereotype threat, see Claude Steele and Joshua Aronson, "Stereotype Threat and the Intellectual Test Performance of African-Americans," *Journal of Personality and Social Psychology* 69 (1995): 797–811. The experiment with African American and white Stanford students is taken from this paper.

51. Nationally, the scores for African American students are improving and the achievement gap between African American and white students has been reduced in the past decade for grades four and eight, but as of 2007, white students still performed at least 26 points higher in all subjects on a test worth 500 points. A. Vanneman et al., *Achievement Gaps: How Black and White Students in Public Schools Perform in Mathematics and Reading on the National Assessment of Educational Progress* (Washington, DC: NCES, 2009).

52. For a meta-analysis on the effects of stereotype threat for people of color and women, see H. H. D. Nguyen and A. M. Ryan, "Does Stereotype Threat Affect Test Performance of Minorities and Women? A Meta-Analysis of Experimental Evidence," *Journal of Applied Psychology* 93 (2008): 1314–34.

53. Von Bakanic, *Prejudice: Attitudes About Race, Class, and Gender* (New York: Pearson, 2008).

54. To dig into the detailed research studies on women's math performance, see M. Inzlicht and T. Ben-Zeev, "A Threatening Intellectual Environment: Why Females Are Susceptible to Experiencing Problem-Solving Deficits in the

Presence of Males," *Psychological Science* 11 (2000): 365–71; see also S. J. Spencer, C. M. Steele, and D. M. Quinn, "Stereotype Threat and Women's Math Performance," *Journal of Experimental Social Psychology* 35 (1999): 4–28.

55. For research showing math-related stereotype threat appearing in elementary -school-age girls in the United States, see Nalini Ambady et al., "Stereotype Susceptibility in Children: Effects of Identity Activation on Quantitative Performance," *Psychological Science* 12, no. 5 (2001): 385–90. For findings of math-related stereotype threat appearing in girls ages eleven to twelve in France, see Pascal Huguet and Isabelle Regner, "Stereotype Threat Among Schoolgirls in Quasi-Ordinary Classroom Circumstances," *Journal of Educational Psychology* 99, no. 3 (2007): 545–60.

56. These findings are taken from one of the classic, most frequently cited works on stereotype threat and gender differences on math test performance, Spencer, Steele, and Quinn, "Stereotype Threat and Women's Math Performance." The particular results that I describe are taken from experiment 3. Women's average test scores actually dropped from 17 to 7, a 58.8 percent drop, but to make the findings easier to understand, I report the score dropping from 10 to 4 (also a 58.8 percent drop). Men's average test scores, by contrast, showed no significant change, going from 18 to 21 on the same test. (That might sound as though men's scores actually improved when they were asked to think how they might be evaluated, but the difference in men's scores was not statistically significant.)

57. Joshua Aronson et al., "When White Men Can't Do Math: Necessary and Sufficient Factors in Stereotype Threat," *Journal of Experimental Social Psychology* 35, no. 1 (1999): 29–46.

58. Stereotype threat on math tests is greatest for women who are highly motivated, as shown in Catherine Good, Joshua Aronson, and Jayne Ann Harder, "Problems in the Pipeline: Stereotype Threat and Women's Achievement in High-Level Math Courses," *Journal of Applied Developmental Psychology* 29, no. 1 (2008): 17–28. These findings have also been demonstrated in the workplace; highly motivated female employees are more likely to experience stereotype threat, as shown in Loriann Roberson and Carol T. Kulik, "Stereotype Threat at Work," *Academy of Management Perspectives* 21 (2007): 24–40.

59. For the original study on the Capilano Bridge, see Donald Dutton and Arthur P. Aron, "Some Evidence for Heightened Sexual Attraction Under Conditions of High Anxiety," *Journal of Personality and Social Psychology* 30 (1974): 510–17. For a more recent overview of how we misattribute our emotions, see B. Keith Payne

et al., "An Inkblot for Attitudes: Affect Misattribution As Implicit Measurement," *Journal of Personality and Social Psychology* 89, no. 3 (2005): 277–93.

60. D. L. Oswald and R. D. Harvey, "Hostile Environments, Stereotype Threat, and Math Performance Among Undergraduate Women," *Current Psychology* 19 (2001): 338–56.

61. For a detailed overview of the many ways that the body shows a stress reaction to stereotype threat — everything from an increased heart rate and pupil dilation to increased activation in brain regions such as the anterior cingulate cortex (evoking similar brain patterns to those found when someone is in physical pain) — see Wendy Berry Mendes and Jeremy Jamieson, "Embodied Stereotype Threat: Exploring Brain and Body Mechanisms Underlying Performance Impairments," in *Stereotype Threat*, 51–68.

62. M. C. Murphy, C. M., Steele, and J. J. Gross, "Signaling Threat: How Situational Cues Affect Women in Math, Science, and Engineering Settings," *Psychological Science* 18, no. 10 (2007): 879–85.

63. See Inzlicht and Ben-Zeev, "A Threatening Intellectual Environment." You might be wondering how the men fared based on the composition of test-takers in the room. It made no difference to them — men scored a consistent average of 67 percent regardless of whether they were surrounded by men or women. Their thinking wasn't diverted because they weren't feeling threatened by any negative labels.

64. Nina Totenberg, "Sandra Day O'Connor's Supreme Legacy: First Female High Court Justice Reflects on 22 Years on Bench," *All Things Considered*, NPR, May 14, 2003, www.npr.org/templates/story/story.php?storyId=1261400.

65. Using the term *critical mass* in this context comes from Claude Steele, *Whistling Vivaldi: How Stereotypes Affect Us and What We Can Do* (New York: W. W. Norton, 2010), 136.

66. I developed these questions using empirical studies of stereotype threat. Most of the questions draw on the work of Courtney von Hippel, a psychologist at the University of Queensland who studies stereotype threat in the workplace and who has developed a measure of stereotype threat for working women. See Courtney von Hippel, Denise Sekaquaptewa, and Matthew McFarlane, "Stereotype Threat Among Women in Finance: Negative Effects on Identity, Workplace Well-Being, and Recruiting," *Psychology of Women Quarterly* 39, no. 3 (September 2015): 405–14, and Courtney von Hippel et al., "Stereotype Threat: Antecedents and Consequences for Working Women," *European Journal of Social Psychology* 41, no. 2 (2011): 151–61. See also Roberson and Kulik, "Stereotype Threat at Work."

67. Crystal L. Hoyt et al., "The Impact of Blatant Stereotype Activation and Group Sex-Composition on Female Leaders," *Leadership Quarterly* 21, no. 5 (2010): 716–32.

68. If you'd like to read a relatively accessible paper on how stereotype threat operates in the workplace, see Roberson and Kulik, "Stereotype Threat at Work." For a more technical account and plenty of intricate details, see Inzlicht and Schmader, *Stereotype Threat*.

69. For a popular definition of working memory, see Scott Barry Kaufman, "In Defense of Working Memory Training," ScientificAmerican.com, April 15, 2013, blogs.scientificamerican.com/beautiful-minds/in-defense-of-working-memory-training/. For a classic definition of working memory and its different components, see A. D. Baddeley and G. Hitch, "Working Memory," in *The Psychology of Learning and Motivation,* ed. G. A. Bower (New York: Academic Press, 1974), 47–89.

70. The ruminations that plague people during stereotype threat are captured in Claude Steele's popular book *Whistling Vivaldi.* For a more technical explanation of these ruminations, see Toni Schmader and Sian Beilock, "An Integration of Processes That Underlie Stereotype Threat," in *Stereotype Threat,* 34–50.

71. For empirical research demonstrating that stereotype threat uses up working memory, see Toni Schmader and Michael Johns, "Converging Evidence That Stereotype Threat Reduces Working Memory Capacity," *Journal of Personality and Social Psychology* 85, no. 3 (2003): 440–52, and Sian Beilock, Robert J. Rydell, and Allen R. McConnell, "Stereotype Threat and Working Memory: Mechanisms, Alleviation, and Spillover," *Journal of Experimental Psychology: General* 136, no. 2 (2007): 256–76. For a broader view of the impact that stereotype threat has on working memory, problem-solving, and reasoning capacity, see Toni Schmader, Michael Johns, and Chad Forbes, "An Integrated Process Model of Stereotype Threat Effects on Performance," *Psychological Review* 115, no. 2 (2008): 336–56.

72. The anti-inflammatory imagery is a fitting analogy; it was proposed by Geoffrey Cohen and his colleagues in G. L. Cohen, V. Purdie-Vaughns, and J. Garcia, "An Identity Threat Perspective on Intervention," in *Stereotype Threat,* 280–96.

73. In fact, "Knowing Is Half the Battle" is the title of the research article that first explored whether learning about stereotype threat helped or hindered women. See Michael Johns, Toni Schmader, and Andy Martens, "Knowing Is Half the Battle: Teaching Stereotype Threat As a Means of Improving Women's Math Performance," *Psychological Science* 16, no. 3 (2005): 175–79.

74. See ibid. for research showing that learning about stereotype threat inoculated women from its effects.

75. Joanne Wood et al., "Positive Self-Statements: Power for Some, Peril for Others," *Psychological Science* 20, no. 7 (2009): 860–66.

76. Claude Steele, the same researcher now famous for his work on stereotype threat, first introduced this values-affirmation activity as a way for smokers to improve their self-image, which can lead them to make healthier choices; see Claude M. Steele, "The Psychology of Self-Affirmation: Sustaining the Integrity of the Self," in *Advances in Experimental Social Psychology*, vol. 21, ed. L. Berkowitz (San Diego: Academic Press, 1988), 261–302. The impacts of different kinds of self-affirmation activities have since been rigorously tested and documented in the laboratory and in real-life settings. For a thorough review of research studies on self-affirmation, see A. McQueen and W.M.P. Klein, "Experimental Manipulations of Self-Affirmation: A Systematic Review," *Self and Identity* 5 (2006): 289–354.

77. This list of possible core values is taken from the affirmation-training study that was conducted by Gregory M. Walton et al., "Two Brief Interventions to Mitigate a 'Chilly Climate' Transform Women's Experience, Relationships, and Achievement in Engineering," *Journal of Educational Psychology* 107, no. 2 (May 2015): 468–85.

78. This explanation of what the self-affirmation activity achieves is taken from David S. Yeager and Gregory M. Walton, "Social-Psychological Interventions in Education: They're Not Magic," *Review of Educational Research* 81, no. 2 (2011): 267–301; quote on page 280.

79. Several researchers have written about social belonging as being the key to values-affirmation activities. For a recent and thorough explanation of this, see Nurit Shnabel et al., "Demystifying Values-Affirmation Interventions: Writing About Social Belonging Is Key to Buffering Against Identity Threat," *Personality and Social Psychology Bulletin* 39, no. 5 (2013): 663–76. Researchers have posited other explanations for how self-affirmation works. Some scientists point to the fact that self-affirmation helps people focus on the problem rather than avoiding it, and some scientists believe self-affirmation is effective because it helps people see that the threat is small relative to life in general. To compare competing theories about why self-affirmation works, see Geoffrey L. Cohen and David K. Sherman, "The Psychology of Change: Self-Affirmation and Social Psychological Intervention," *Annual Review of Psychology* 65 (2014): 333–71.

80. The saying-is-believing notion was explored in Gregory M. Walton and

Geoffrey L. Cohen, "A Brief Social-Belonging Intervention Improves Academic and Health Outcomes of Minority Students," *Science* 331 (2011): 1447–51.

81. In most studies that use self-affirmation to address stereotype threat, people write about values that have little to do with the challenge they're about to face. A woman who is about to take a math test can write about why she values her friendships, and that improves her performance. An African American office worker can write about why he values spending time in nature, and that improves his ability to receive feedback from his boss.

82. Most of the research on the benefits of self-affirmation has been done to eliminate the academic-achievement gaps between the races or math-performance gaps between the genders. On the issue of race, see G. L. Cohen et al., "Reducing the Racial Achievement Gap: A Social-Psychological Intervention," *Science* 313 (2006): 1307–10. For work on how self-affirmation boosts women's math performance, see A. Martens et al., "Combating Stereotype Threat: The Effect of Self-Affirmation on Women's Intellectual Performance," *Journal of Experimental Social Psychology* 42 (2006): 236–43.

83. The benefits of self-affirmation have been documented for many intellectual skills. For research on how self-affirmation improves problem-solving, see J. David Creswell et al., "Self-Affirmation Improves Problem-Solving Under Stress," *PLOS ONE* 8, no. 5 (2013): e62593, doi:10.1371/journal.pone.0062593. For research on how self-affirmation improves interest in information that people usually avoid when they're making decisions, see J. L. Howell and J. A. Shepperd, "Reducing Information Avoidance Through Affirmation," *Psychological Science* 23, no. 2 (2012): 141–45.

84. Niro Sivanathan et al., "The Promise and Peril of Self-Affirmation in De-Escalation of Commitment," *Organizational Behavior and Human Decision Processes* 107, no. 1 (2008): 1–14.

3. Hello, Risk-Taker

1. David A. Kaplan, *The Silicon Boys: And Their Valley of Dreams* (New York: William Morrow Paperbacks, 2000), 193.

2. Vivienne isn't thrilled with the term *transgender* even though it is generally accepted language. She prefers to say that she has undergone gender transition, much like other people have survived cancer. She identifies as female and a woman rather than as transgender or a transwoman.

3. I should also note that most venture capitalists, more than 90 percent, are men. A study done at Babson College found that women made up 6 percent of all venture capitalists in 2014, down from 10 percent in 1999; see Candida Brush et al., *Women Entrepreneurs 2014: Bridging the Gender Gap in Venture Capital*

(Babson Park, MA: Babson College, 2014). A 2014 study by *Fortune* magazine found that at the senior-partner level, where the big decisions are made, only 4.2 percent of executives were women; see Dan Primack, "Venture Capital's Stunning Lack of Female Decision-Makers," *Fortune*, February 6, 2014.

4. Alison W. Brooks et al., "Investors Prefer Entrepreneurial Ventures Pitched by Attractive Men," *Proceedings of the National Academy of Sciences* 111, no. 12 (2014): 4427–31.

5. Brooks isn't the only one to report that investors prefer to put their dollars behind men. Sarah Thébaud, a professor at the University of California, Santa Barbara, found similar results in a study in which she had participants evaluate the profitability and likelihood of investment for the same pitch, one made by a male entrepreneur, one made by a female; see Sarah Thébaud, "Status Beliefs and the Spirit of Capitalism: Accounting for Gender Biases in Entrepreneurship and Innovation," *Social Forces* (2015), doi:10.1093/sf/sov042. When the proposal wasn't innovative, then investors were much more interested in the pitches made by men, just as Brooks found. But when the proposed startup was truly innovative, something that the investor had never heard of before, it seemed to be a signal that the women were exceptional, that they defied the stereotypes, and investors were more willing to take a risk on the women. So there is hope, but it's a high bar if women need to propose ideas that are more innovative than men's.

6. These are the results from a second study conducted by Alison Brooks and her colleagues. In this study, participants acted as investors and picked one of two veterinary-technology projects to fund. The investors didn't see the entrepreneurs, but they watched a video of images related to the idea being presented, and they heard a voice-over narration of the pitch. These voice-overs were done by either a man or a woman and were as identical in content and tone as possible, but 68.33 percent of the participants chose to fund the project pitched by a man and only 31.7 percent chose to fund the project pitched by a woman.

7. Brush et al., *Women Entrepreneurs 2014*. For the 2014 Diana Project, as it's called, the researchers looked at a database of 6,793 U.S. companies that received venture capital funds from 2011 to 2013.

8. Bernd Figner and Elke U. Weber, "Who Takes Risks When and Why? Determinants of Risk Taking," *Current Directions in Psychological Science* 20, no. 4 (2011): 211–16.

9. Yaniv Hanoch, Joseph G. Johnson, and Andreas Wilke, "Domain Specificity in Experimental Measures and Participant Recruitment: An Application to Risk-Taking Behavior," *Psychological Science* 17, no. 4 (2006): 300–304.

10. The exact origin of the word *wuss* isn't clear, but the speculation among language scholars is that it first appeared in the United States in the 1980s and reflects a blend of these two insulting words. See David Crystal, "Keep Your English Up-to-Date: Wuss," *BBC Learning English*, 2005, accessed October 21, 2014, downloads.bbc.co.uk/worldservice/learningenglish/uptodate/pdf/uptodate2_wuss_transcript_070316.pdf.

11. Barbara A. Morrongiello and Theresa Dawber, "Parental Influences on Toddlers' Injury-Risk Behaviors: Are Sons and Daughters Socialized Differently?," *Journal of Applied Developmental Psychology* 20, no. 2 (1999): 227–51. See also B. A. Morrongiello, D. Zdzieborski, and J. Norman, "Understanding Gender Differences in Children's Risk-Taking and Injury: A Comparison of Mothers' and Fathers' Reactions to Sons and Daughters Misbehaving in Ways That Lead to Injury," *Journal of Applied Developmental Psychology* 31 (2010): 322–29.

12. Sheryl Ball, Catherine C. Eckel, and Maria Heracleous, "Risk Aversion and Physical Prowess: Prediction, Choice and Bias," *Journal of Risk and Uncertainty* 41, no. 3 (2010): 167–93.

13. For studies where men see women as much more risk-averse than women actually are, see Catherine C. Eckel and Philip J. Grossman, "Sex Differences and Statistical Stereotyping in Attitudes Toward Financial Risk," *Evolution and Human Behavior* 23, no. 4 (2002): 281–95; Dinky Daruvala, "Gender, Risk and Stereotypes," *Journal of Risk and Uncertainty* 35, no. 3 (2007): 265–83; and Catherine C. Eckel and Philip J. Grossman, "Forecasting Risk Attitudes: An Experimental Study Using Actual and Forecast Gamble Choices," *Journal of Economic Behavior and Organization* 68, no. 1 (2008): 1–17.

14. Philip J. Grossman, "Holding Fast: The Persistence and Dominance of Gender Stereotypes," *Economic Inquiry* 51, no. 1 (2013): 747–63.

15. Deborah A. Prentice and Erica Carranza, "What Women and Men Should Be, Shouldn't Be, Are Allowed to Be, and Don't Have to Be: The Contents of Prescriptive Gender Stereotypes," *Psychology of Women Quarterly* 26, no. 4 (2002): 269–81. To calculate these rankings, I used table 2 for women and table 3 for men. The items in each table are ordered from most desirable to least desirable. To determine where "willing to take risks" ranked for men, I started at the top of the table for men and counted down until I reached that quality; "willing to take risks" was fourteenth. I used the same process to determine where "willing to take risks" fell for women, counting through what the authors call "Intensified prescriptions" then moving on to the "Relaxed prescriptions."

16. This list of risk-takers is featured in Deborah Perry Piscione's book on risk-

taking, *The Risk Factor* (New York: Macmillan, 2014). She describes them as a "new crop of leaders" who "understand that willingness to take risks is at the very heart of their leadership" (page 45). Piscione was a media commentator on CNN, CNBC, MNNBC, ABC, Fox News, and PBS, and although she's just one example, she's one of many journalists to spotlight these men as risk-takers that the rest of the business world emulates.

17. Not all Walgreens stores have a Theranos. At the time I write this, the Theranos labs are common in the Walgreens in Phoenix, Arizona; see Ron Leuty, "Theranos Sticks It to Critics, Plans Expansion of Lab Services," *San Francisco Business Times*, May 5, 2015.

18. The details on Elizabeth Holmes and her remarkable story come largely from an extensive and fascinating piece about Holmes and Theranos written by Roger Parloff, "This CEO Is Out for Blood," *Fortune,* June 12, 2014, fortune.com/2014/06/12/theranos-blood-holmes/. Other details came from Rachel Crane, "She's America's Youngest Female Billionaire — and a Dropout," *CNN Money*, October 16, 2014, money.cnn.com/2014/10/16/technology/theranos-elizabeth-holmes/.

19. Victoria L. Brescoll, Erica Dawson, and Eric L. Uhlmann, "Hard Won and Easily Lost: The Fragile Status of Leaders in Gender-Stereotype-Incongruent Occupations," *Psychological Science* 21, no. 11 (2010): 1640–42.

20. Brescoll and her team conducted a separate study to identify occupations that were strongly associated with women and another set of occupations strongly associated with men. President of a women's college and a police chief were rated as equally high status but strongly gender-stereotyped.

21. Victoria Brescoll, personal communication with the author, June 17, 2015.

22. These are just some of the reasons for the glass ceiling that the Feminist Majority Foundation lists in its article, "Empowering Women in Business," www.feminist.org/research/business/ewb_glass.html.

23. Denise R. Beike and Travis S. Crone, "When Experienced Regret Refuses to Fade: Regrets of Action and Attempting to Forget Open Life Regrets," *Journal of Experimental Social Psychology* 44, no. 6 (2008): 1545–50.

24. Nina Hattiangadi, Victoria Husted Medvec, and Thomas Gilovich, "Failing to Act: Regrets of Terman's Geniuses," *International Journal of Aging and Human Development* 40, no. 3 (1995): 175–85. The finding that people regret their inactions more than their actions is common in the regret literature. For a delightfully written overview of how we regret the risks we don't take, see Daniel Gilbert, *Stumbling on Happiness* (New York: Vintage Books, 2006).

25. Herminia Ibarra and Otilia Obodaru, "Women and the Vision Thing," *Harvard Business Review* 87, no. 1 (2009): 65.

26. *Women Leaders: Research Paper* (Princeton, NJ: Caliper Corporation, December 2014), https://www.calipercorp.com/home-3/banner-women -leaders-white-paper/.

27. Raymond S. Nickerson, "Confirmation Bias: A Ubiquitous Phenomenon in Many Guises," *Review of General Psychology* 2, no. 2 (June 1998): 175–220.

28. Julie A. Nelson, "The Power of Stereotyping and Confirmation Bias to Overwhelm Accurate Assessment: The Case of Economics, Gender, and Risk Aversion," *Journal of Economic Methodology* 21, no. 3 (2014): 211–31. See also Julie A. Nelson, "Not-So-Strong Evidence for Gender Differences in Risk-Taking," *Feminist Economics* (July 2015).

29. Julie Nelson, personal communication with the author, September 18, 2014.

30. Nancy M. Carter and Christine Silva, "The Myth of the Ideal Worker: Does Doing All the Right Things Really Get Women Ahead?," Catalyst, 2011, http:// www.catalyst.org/knowledge/myth-ideal-worker-does-doing-all-right-things -really-get-women-ahead.

31. Joan C. Williams and Rachel Dempsey, *What Works for Women at Work* (New York: New York University Press, 2014).

32. For data on how women CEOs are older and have more education and job experience than male CEOs, see Jeremy Donovan, *Women Fortune 500 CEOs: Held to Higher Standards* (New York: American Management Association, 2015). For data on women being at their companies a long time before they become CEOs, see Sarah Dillard and Valerie Lipschitz, "Research: How Female CEOs Actually Get to the Top," *Harvard Business Review*, November 6, 2014.

33. Kimberly A. Daubman and Harold Sigall, "Gender Differences in Perceptions of How Others Are Affected by Self-Disclosure of Achievement," *Sex Roles* 37, nos. 1–2 (1997): 73–89.

34. Laurie A. Rudman, "Self-Promotion As a Risk Factor for Women: The Costs and Benefits of Counterstereotypical Impression Management," *Journal of Personality and Social Psychology* 74, no. 3 (1998): 629–45. See also Laurie A. Rudman and Peter Glick, "Feminized Management and Backlash Toward Agentic Women: The Hidden Costs to Women of a Kinder, Gentler Image of Middle Managers," *Journal of Personality and Social Psychology* 77, no. 5 (1999): 1004–10.

35. James P. Byrnes, David C. Miller, and William D. Schafer, "Gender Differences in Risk Taking: A Meta-Analysis," *Psychological Bulletin* 125, no. 3 (1999): 367. As of January 2016, this article had been cited by 1,688 sources, according to Google Scholar, making it what I imagine is the most cited work on gender and risk-taking to date.

36. The researcher who made this simple but brilliant observation that only 60

percent of the studies show men taking more risks was Julie Nelson in her 2014 paper "Are Women Really More Risk-Averse Than Men? A Re-Analysis of the Literature Using Expanded Methods," *Journal of Economic Surveys* 29, no. 3 (May 2014): 566–85.

37. Christine R. Harris, Michael Jenkins, and Dale Glaser, "Gender Differences in Risk Assessment: Why Do Women Take Fewer Risks Than Men?," *Judgment and Decision Making* 1, no. 1 (2006): 48–63.

38. Elke U. Weber, Ann-Renee Blais, and Nancy E. Betz, "A Domain-Specific Risk-Attitude Scale: Measuring Risk Perceptions and Risk Behaviors," *Journal of Behavioral Decision Making* 15, no. 4 (2002): 263–90. The items I've listed are taken from the appendices in the article.

39. For research on women restarting their careers, see Figner and Weber, "Who Takes Risks When and Why?" For research on self-disclosure, see Kathryn Dindia and Mike Allen, "Sex Differences in Self-Disclosure: A Meta-Analysis," *Psychological Bulletin* 112, no. 1 (1992): 106. See also A. J. Rose and K. D. Rudolph, "A Review of Sex Differences in Peer Relationship Processes: Potential Trade-Offs for the Emotional and Behavioral Development of Girls and Boys," *Psychological Bulletin* 132 (2006): 89–131.

40. Christopher F. Karpowitz and Tali Mendelberg, *The Silent Sex: Gender, Deliberation, and Institutions* (Princeton, NJ: Princeton University Press, 2014), 143.

41. In 2013, 77 percent of all single parents in the United States were women; see U.S. Census Bureau, table f1: Family Households, by Type, Age of Own Children, Age of Family Members, and Age, Race and Hispanic Origin of Householder, www.census.gov/hhes/families/data/cps2013F.html. Data showing that single parents receive less education than married parents can be found at the Pew Research Center website; see Gretchen Livingston, "The Links Between Education, Marriage and Parenting," Pew Research Center, November 27, 2013, www.pewresearch.org/fact-tank/2013/11/27/the-links -between-education-marriage-and-parenting/.

42. A fascinating article on single-parent households across developed nations was written by Catherine Rampell, "Single Parents, Around the World," *New York Times*, March 10, 2010. If you're interested in the source data on single-parent households, it's available through the OECD Family Database, www.oecd.org /social/family/41919559.pdf.

43. For gender differences in alcohol, tobacco, and drug use, see Louisa Degenhardt et al., "Toward a Global View of Alcohol, Tobacco, Cannabis, and Cocaine Use: Findings from the WHO World Mental Health Surveys," *PLoS Medicine* 5, no. 7 (2008): e141. For gender differences in drowning and other

accidents, see I. Waldron, C. McCloskey, and I. Earle, "Trends in Gender Differences in Accident Mortality: Relationships to Changing Gender Roles and Other Societal Trends," *Demographic Research* 13 (2005): 415–54. For gender differences in reckless driving, see G. Beattie, "Sex Differences in Driving and Insurance Risk: Understanding the Neurobiological and Evolutionary Foundations of the Differences," Social Issues Research Centre (Manchester, UK: University of Manchester, 2008). For gender differences in extreme sports, see Victoria Robinson, *Everyday Masculinities and Extreme Sport: Male Identity and Rock Climbing* (London: Bloomsbury Academic, 2008), and Harris, Jenkins, and Glaser, "Gender Differences in Risk Assessment."

44. Figner and Weber, "Who Takes Risks When and Why?"

45. Gary Charness and Uri Gneezy, "Strong Evidence for Gender Differences in Risk Taking," *Journal of Economic Behavior and Organization* 83, no. 1 (2012): 50–58. Another survey of the literature that is often cited as concluding that women are more risk-averse while men are more risk-taking is Rachel Croson and Uri Gneezy, "Gender Differences in Preferences," *Journal of Economic Literature* (2009): 448–74.

46. Nelson, "Are Women Really More Risk-Averse Than Men?"

47. Julie Nelson was kind enough to spell out these statistics for me in a personal phone call on September 18, 2014, and in a personal e-mail on September 21, 2014.

48. This particular finding of men paying 2.2 times more for the gamble is reported in an unpublished work by Alice Wieland and Rakesh Sarin, "Gender Differences in Risk Aversion: A Theory of When and Why," 2012. Variations on the task itself, though, are common in economic and statistical research; see Norman Lloyd Johnson and Samuel Kotz, *Urn Models and Their Application: An Approach to Modern Discrete Probability Theory* (New York: Wiley, 1977), 402, or, more recently, Lex Borghans et al., "Gender Differences in Risk Aversion and Ambiguity Aversion," *Journal of the European Economic Association* 7, nos. 2–3 (2009): 649–58.

49. Johnnie E. V. Johnson and Philip L. Powell, "Decision Making, Risk and Gender: Are Managers Different?," *British Journal of Management* 5, no. 2 (1994): 123–38.

50. Peggy D. Dwyer, James H. Gilkeson, and John A. List, "Gender Differences in Revealed Risk Taking: Evidence from Mutual Fund Investors," *Economics Letters* 76, no. 2 (2002): 151–58.

51. Anna Dreber et al., "Dopamine and Risk Choices in Different Domains:

Findings Among Serious Tournament Bridge Players," *Journal of Risk and Uncertainty* 43, no. 1 (2011): 19–38.

52. Carol Tavris and Elliot Aronson, *Mistakes Were Made (but Not by Me)*, rev. ed. (Boston: Houghton Mifflin Harcourt, 2015).

53. Allison E. Seitchik, Jeremy Jamieson, and Stephen G. Harkins, "Reading Between the Lines: Subtle Stereotype Threat Cues Can Motivate Performance," *Social Influence* 9, no. 1 (2014): 52–68.

54. Priyanka B. Carr and Claude M. Steele, "Stereotype Threat Affects Financial Decision Making," *Psychological Science* 21, no. 10 (2010): 1411–16. The findings that I'm reporting are from experiment 2. Carr and Steele discuss all of their findings in terms of the number of times people went with the low-risk option. In the math group, men went for the low-risk, relatively sure gambles approximately 3.8 out of 14 times (27 percent) and women went for the low-risk, relatively sure gambles approximately 8.5 out of 14 times (61 percent), so women went for the relatively sure thing twice as often. This difference in the math condition was statistically significant. In the puzzle-solving group, men went for the low-risk option approximately 6.5 out of 14 times (46 percent) and women went for the low-risk option approximately 6 out of 14 times (43 percent), and the difference was not statistically significant.

55. At least, most tests of decision-making that are looking at gender begin this way. If the researchers aren't interested in sex differences, they might skip the "Tell us your sex" part.

56. Jonathan R. Weaver, Joseph A. Vandello, and Jennifer K. Bosson, "Intrepid, Imprudent, or Impetuous? The Effects of Gender Threats on Men's Financial Decisions," *Psychology of Men and Masculinity* 14, no. 2 (2013): 184.

57. Jonathan Weaver, personal communication with the author, June 12, 2015.

58. Although the men were all heterosexual, it is worth noting that the group of men Weaver and his colleagues tested was relatively diverse on other dimensions, ranging in age from eighteen to forty, representing several different races and ethnicity groups (39.5 percent white, 28.9 percent black, 15.8 percent Hispanic/Latino, 7.9 percent biracial, 2.6 percent Arabic / Middle Eastern, 2.6 percent Asian, and 2.6 percent other).

59. Jonathan Weaver, personal communication with the author, June 11, 2015. But for over twenty years, social scientists have been talking about how manhood and masculinity must be continually affirmed and that men feel they are under constant threat of being exposed as not being "real men." See John Stoltenberg, *The End of Manhood: A Book for Men of Conscience* (New York: Penguin, 1994), and Michael Kimmel, "Masculinity as Homophobia: Fear, Shame, and

Silence in the Construction of Gender Identity," in *Theorizing Masculinities*, eds. Harry Broad and Michael Kaufman (Thousand Oaks, CA: Sage, 1994), 119–41.

60. Joseph A. Vandello et al., "Precarious Manhood," *Journal of Personality and Social Psychology* 95, no. 6 (2008): 1325–39.

61. Joelle Emerson, personal communication with the author, February 10, 2015.

62. Kieran Snyder, e-mail to the author, March 24, 2015.

63. Kaplan, *The Silicon Boys*.

64. These findings are taken from three different studies: James Flynn, Paul Slovic, and Chris K. Mertz, "Gender, Race, and Perception of Environmental Health Risks," *Risk Analysis* 14, no. 6. (1994): 1101–8; Melissa L. Finucane et al., "Gender, Race, and Perceived Risk: The 'White Male' Effect," *Health, Risk and Society* 2, no. 2 (2000): 159–72; and Dan M. Kahan et al., "Culture and Identity-Protective Cognition: Explaining the White-Male Effect in Risk Perception," *Journal of Empirical Legal Studies* 4, no. 3 (2007): 465–505.

65. Flynn, Slovic, and Mertz, "Gender, Race, and Perception."

66. Finucane et al., "Gender, Race, and Perceived Risk." You might be wondering about racial or ethnic differences among women. In their telephone survey conducted in 1997 and 1998, Finucane and her colleagues found that Hispanic women in the United States saw the world as slightly more risky than any other demographic group they included. Their graph on page 133 shows the risk ratings of white men on the farthest left side of the graph and Hispanic women's ratings on the farthest right side of the graph, with white men believing that blood transfusions, handguns, and nuclear power plants pose little to no risk but Hispanic women seeing these same entities as posing moderate to high risks. But when Hispanic, Asian American, and African American women were combined into a single group of "nonwhite" women, their perceptions of risk clustered together with both white women and nonwhite men. It was the white males who stood out as seeing the world as safer than everyone else.

67. Dan Kahan, at Yale Law School, and his colleagues provide an in-depth analysis of the source of the white-male effect. They see risk perception as a worldview. If you believe in a hierarchical society, one in which only a few select individuals can get ahead, and you believe that every person needs to fend for himself to attain one of those select spots at the top, then you're highly motivated to be skeptical of risks. You need to be. You need to see the world as your oyster, as full of things within your control if you try hard enough. See Kahan et al., "Culture and Identity-Protective Cognition."

68. Flynn, Slovic, and Mertz, "Gender, Race, and Perception," 1107. I thank Julie

Nelson at the University of Massachusetts, Boston, for bringing this quote to my attention.

69. And yes, these are real questions from dating websites. As of March 2015, Match.com asks users about their income, eHarmony asks users to rate how well the sentence "I feel unable to deal with things" describes them, and OkCupid asks the question about starving animals or children. It's possible to skip these questions, but they are included.

70. Aaron Smith and Maeve Duggan, "Online Dating and Relationships," Pew Research Center (October 21, 2013), pewinternet.org/Reports/2013/Online-Dating.aspx. Quote from page 5.

71. Data on Siren's absence of harassment is based on CEO and founder Susie J. Lee's e-mail to the author on July 30, 2015.

72. When Susie and I spoke, Siren was limited to Seattle. They are planning to go national, but there are apps in other parts of the United States that also give women more control of the experience, such as Wyldfire and Bumble. Dating websites and applications have different ratios of women to men. As of 2012, 71.8 percent of the users on Chemistry.com were women, as were 68.6 percent of the dating profiles on eHarmony. Some websites, such as Match.com, have a more balanced gender ratio, with 55 percent women users. See Online Dating Demographics, WebPersonalsOnline.com, 2012, accessed August 24, 2015, www.webpersonalsonline.com/demographics_online_dating.html.

73. Susan R. Fisk, "Risky Spaces, Gendered Places: How Intersecting Beliefs About Gender and Risk Reinforce and Recreate Gender Inequality" (doctoral dissertation, Stanford University, 2015). Readers who are well acquainted with the gender literature might be wondering if I've misspelled Susan's last name. More people know the Susan Fiske at Princeton University, who is an established psychology professor who studies how stereotyping and discrimination shrink or flourish based on our social interactions, but there is also a Susan Fisk (no *e*) at Kent State University who is earlier in her career. Fisk also studies gender and stereotyping, which makes things a little confusing, but she specifically looks at how social pressures affect men and women's experiences of risk. Both researchers do important work, but I'm describing the work of the younger Susan here.

74. Susan Fisk, personal communication with the author, August 2, 2015.

75. Suzy Welch, *10-10-10: 10 Minutes, 10 Months, 10 Years* (New York: Scribner, 2009).

76. The instructions for the premortem are taken from Daniel Kahneman, *Thinking, Fast and Slow* (New York: Farrar, Straus and Giroux, 2011), 264.

77. The original research study was conducted by Deborah J. Mitchell, J. Edward

Russo, and Nancy Pennington, "Back to the Future: Temporal Perspective in the Explanation of Events," *Journal of Behavioral Decision Making* 2, no. 1 (1989): 25–38. The president scenario is a variation on a scenario given by J. Edward Russo and Paul J. H. Schoemaker, *Winning Decisions: Getting It Right the First Time* (New York: Crown Business, 2002), 111–12.

78. Jeremy A. Yip and Stéphane Côté, "The Emotionally Intelligent Decision-Maker: Emotion-Understanding Ability Reduces the Effect of Incidental Anxiety on Risk-Taking," *Psychological Science* 24, no. 1 (2013): 48–55.

79. Chip Heath and Dan Heath discuss tripwires in a most entertaining way in their book *Decisive: How to Make Better Choices in Life and Work* (New York: Crown Business, 2013).

4. Women's Confidence Advantage

1. Jodi Kantor, "Harvard Business School Case Study: Gender Equity," *New York Times,* September 7, 2013, www.nytimes.com/2013/09/08/education/harvard-case-study-gender-equity.html?pagewanted=alland_r=0.

2. In 2010, women made up 40 percent of the women in the MBA program yet were only 20 percent of the Baker Scholars. See Laura Ratcliff, "Next Generation of Female Leaders Needs Strong Mentors," *Glass Hammer,* www.theglasshammer.com/news/2011/05/25/next-generation-of-female-leaders-need-strong-mentors/.

3. Nanette Fondas, "First Step to Fixing Gender Bias in Business School: Admit the Problem," *Atlantic,* September 17, 2013, www.theatlantic.com/education/archive/2013/09/first-step-to-fixing-gender-bias-in-business-school-admit-the-problem/279740/.

4. In a talk that Dean Nohria gave to HBS alums in San Francisco in 2014, he said women "felt disrespected, left out, and unloved by the school. I'm sorry on behalf of the business school . . . The school owed you better, and I promise it will be better." See John Byrne, "Harvard B-School Dean Offers Unusual Apology," Fortune.com, January 29, 2014, fortune.com/2014/01/29/harvard-b-school-dean-offers-unusual-apology/.

5. I interviewed three women who had attended Harvard Business School while the class-participation seminars were offered. The three women represented two different graduating classes, and for confidentiality reasons, I have not used their real names.

6. Kantor, "Harvard Business School Case Study." See also *Annual 2013: A Year in Review* (Cambridge, MA: Harvard Business School, 2013), 11.

7. J. Edward Russo and Paul J. H. Schoemaker, "Managing Overconfidence," *Sloan Management Review* 33, no. 2 (1992): 7–17.

8. Pascal Mamassian, "Overconfidence in an Objective Anticipatory Motor Task," *Psychological Science* 19, no. 6 (2008): 601–6. For one of the earliest research definitions of overconfidence, see S. Oskamp, "Overconfidence in Case-Study Judgments," *Journal of Consulting Psychology* 29 (1965): 261–65.

9. Jonathon D. Brown, "Understanding the Better Than Average Effect: Motives (Still) Matter," *Personality and Social Psychology Bulletin* 38, no. 2 (2012): 209–19.

10. Most people can eat two saltine or soda crackers in one minute without any water. Small but mighty, two crackers are typically enough to soak up all the saliva in your mouth, so beyond two crackers, most of us begin to struggle. People often set a challenge of trying to eat as many as five or six crackers in one minute — after all, they're just crackers — but they are often shocked to find that the task is harder than it seems. See Philippa Wingate and David Woodroffe, *The Family Book: Amazing Things to Do Together* (New York: Scholastic, 2008), 160. For research on how much people know about familiar objects such as bicycles, see Rebecca Lawson, "The Science of Cycology: Failures to Understand How Everyday Objects Work," *Memory and Cognition* 34, no. 8 (2006): 1667–75.

11. Mark D. Alicke and Olesya Govorun, "The Better-Than-Average Effect," in *The Self in Social Judgment*, eds. M. D. Alicke, D. A. Dunning, and J. I. Krueger (New York: Psychology Press, 2005).

12. Mary A. Lundeberg, Paul W. Fox, and Judith Punćcohař, "Highly Confident but Wrong: Gender Differences and Similarities in Confidence Judgments," *Journal of Educational Psychology* 86, no. 1 (1994): 114–21.

13. The gender difference in overconfidence has been demonstrated in a number of areas. One of the classic studies that's cited most often shows that men trade 45 percent more stocks than women; see B. M. Barber and T. Odean, "Boys Will Be Boys: Gender, Overconfidence, and Common Stock Investment," *Quarterly Journal of Economics* 116, no. 1 (2001): 261–92. For a study of gender differences in students' confidence on each individual test item in a psychology course, see Lundeberg, Fox, and Punćcohař, "Highly Confident but Wrong," 114. In the latter study, both men and women were overconfident, but men were especially overconfident on the items they answered incorrectly.

14. Average intelligence is defined as having an IQ score between 90 and 110. Above average is above 110 and below average is below 90.

15. Christopher Chabris and Daniel Simons, *The Invisible Gorilla* (New York: Crown, 2010). Also see Sophie Von Stumm, Tomas Chamorro-Premuzic, and Adrian Furnham, "Decomposing Self-Estimates of Intelligence: Structure and Sex Differences Across 12 Nations," *British Journal of Psychology* 100, no. 2 (2009): 429–42.

16. For an extensive cross-cultural review of the literature on how men and women assess their general intelligence, see Adrian Furnham, "Self-Estimates of Intelligence: Culture and Gender Difference in Self and Other Estimates of Both General (g) and Multiple Intelligences," *Personality and Individual Differences* 31 (2001): 1381–405. For a more recent analysis of twelve nations, see Von Stumm, Chamorro-Premuzic, and Furnham, "Decomposing Self-Estimates of Intelligence." When men and women are asked about specific components of their intelligence, women tend to give higher estimates for their emotional intelligence (EQ) whereas men tend to give higher scores for their overall intelligence (IQ) as measured by tests of spatial and verbal abilities. For comparisons of IQ and EQ, see K. V. Petrides, Adrian Furnham, and G. Neil Matin, "Estimates of Emotional and Psychometric Intelligence: Evidence for Gender-Based Stereotypes," *Journal of Social Psychology* 144, no. 2 (2004): 149–62.

17. Kevin V. Petrides and Adrian Furnham, "Gender Differences in Measured and Self-Estimated Trait Emotional Intelligence," *Sex Roles* 42, nos. 5–6 (2000): 449–61.

18. Sylvia Beyer and Edwin M. Bowden, "Gender Differences in Self-Perceptions: Convergent Evidence from Three Measures of Accuracy and Bias," *Personality and Social Psychology Bulletin* 23, no. 2 (1997): 157–72.

19. For research showing that women highly knowledgeable in finance underestimated their abilities, see Matthias Gysler and Jamie Brown Kruse, *Ambiguity and Gender Differences in Financial Decision Making: An Experimental Examination of Competence and Confidence Effects* (Swiss Federal Institute of Technology, Center for Economic Research, 2002).

20. Marc A. Brackett et al., "Relating Emotional Abilities to Social Functioning: A Comparison of Self-Report and Performance Measures of Emotional Intelligence," *Journal of Personality and Social Psychology* 91, no. 4 (2006): 780–95.

21. For the finding on games of chance, see Anthony Patt, "Understanding Uncertainty: Forecasting Seasonal Climate Change for Farmers in Zimbabwe," *Risk Decision and Policy 6*, no. 2 (2001): 105–19. For the finding on stock trading, see Barber and Odean, "Boys Will Be Boys." For research on overconfidence in driving at night, see John A. Brabyn et al., "Night Driving Self-Restriction: Vision Function and Gender Differences," *Optometry and Vision Science* 82, no. 8 (2005): 755–64.

22. Albert E. Mannes and Don A. Moore, "A Behavioral Demonstration of Overconfidence in Judgment," *Psychological Science* 24, no. 7 (2013): 1190–97.

Pittsburgh has hot, humid summers, and wet, snowy winters, so the temperatures vary widely.

23. For a general review of gender stereotypes in the workplace, see Madeline E. Heilman, "Gender Stereotypes and Workplace Bias," *Research in Organizational Behavior* 32 (2012): 113–35.

24. Samantha C. Paustian-Underdahl, Lisa S. Walker, and David J. Woehr, "Gender and Perceptions of Leadership Effectiveness: A Meta-Analysis of Contextual Moderators," *Journal of Applied Psychology* 99, no. 6 (November 2014): 1129–45.

25. Anne M. Koenig et al., "Are Leader Stereotypes Masculine? A Meta-Analysis of Three Research Paradigms," *Psychological Bulletin* 137, no. 4 (2011): 616–42.

26. Linda Babcock and Sara Laschever, *Women Don't Ask: The High Cost of Avoiding Negotiation — and Positive Strategies for Change* (New York: Bantam Books, 2007).

27. Georges Desvaux, Sandrine Devillard-Hoellinger, and Mary C. Meaney, "A Business Case for Women," *McKinsey Quarterly* (September 2008), www .talentnaardetop.nl/uploaded_files/document/2008_A_business_case_for _women.pdf.

28. Michael Roberto, "Lessons from Everest: The Interaction of Cognitive Bias, Psychological Safety, and System Complexity," *California Management Review* 45, no. 1 (2002): 136–58. These quotes are offered on page 142. It's interesting to note that both of the leaders on the fatal 1996 climb were men. Then again, mountaineering expeditions are typically led by and sought out by men.

29. "The Day the Sky Fell on Everest," *New Scientist* 2449 (May 29, 2004): 15, www .newscientist.com/article/mg18224492.200-the-day-the-sky-fell-on-everest .html.

30. Peter Goodspeed, "Nuclear Hubris Played a Role in Japanese Disaster," *National Post,* March 14, 2011, news.nationalpost.com/full-comment/peter -goodspeed-nuclear-hubris-played-a-role-in-japanese-disaster; Peter Elkind, David Whitford, and Doris Burke, "BP: 'An Accident Waiting to Happen,'" *Fortune,* January 24, 2011, fortune.com/2011/01/24/bp-an-accident-waiting-to -happen/.

31. Michael Corkery, "Meet a Citigroup Whistleblower: Richard M. Bowen III," *Deal Journal,* April 7, 2010, http://blogs.wsj.com/deals/2010/04/07/meet-a -citigroup-whistleblower-richard-m-bowen-iii/.

32. "Prosecuting Wall Street," *60 Minutes,* CBS. On the attempts to keep Bowen quiet, see William D. Cohan, "Was This Whistle-Blower Muzzled?," *New York Times,* September 21, 2013.

33. The original research on the premiums that CEOs paid was first captured in

Mathew L. A. Hayward and Donald C. Hambrick, "Explaining the Premiums Paid for Large Acquisitions: Evidence of CEO Hubris," *Administrative Science Quarterly* 42 (1997): 103–27. It's compellingly described by Chip and Dan Heath in their book *Decisive: How to Make Better Decisions in Life and Work* (New York: Crown Business, 2013).

34. Michael G. Aamodt and Heather Custer, "Who Can Best Catch a Liar? A Meta-Analysis of Individual Differences in Detecting Deception," *Forensic Examiner* 15, no. 1 (2006): 6–11.

35. B. L. Cutler and S. D. Penrod, "Forensically Relevant Moderators of the Relation Between Eyewitness Identification Accuracy and Confidence," *Journal of Applied Psychology* 74, no. 4 (1989): 650. For a fantastic examination of the relationship between confidence and eyewitness testimony, see Chabris and Simons, *The Invisible Gorilla*. Chapter 3 provides a memorable look at the illusions of confidence.

36. Philip E. Tetlock, *Expert Political Judgment: How Good Is It? How Can We Know?* (Princeton, NJ: Princeton University Press, 2005), 233.

37. Daniel Kahneman, *Thinking, Fast and Slow* (New York: Farrar, Straus and Giroux, 2011), 87.

38. Helen Lerner, *The Confidence Myth* (Oakland, CA: Berrett-Koehler, 2015). Data is reported in appendix B. Of the 535 survey respondents, 95.1 percent self-identified as female, giving a total of 509 females.

39. Mariëlle Stel et al., "Lowering the Pitch of Your Voice Makes You Feel More Powerful and Think More Abstractly," *Social Psychological and Personality Science* 3, no. 4 (2012): 497–502.

40. CEOs with deeper voices make more money and they oversee larger corporations; see William J. Mayew, Christopher A. Parsons, and Mohan Venkatachalam, "Voice Pitch and the Labor Market Success of Male Chief Executive Officers," *Evolution and Human Behavior* 34, no. 4 (2013): 243–48. For research on how men with deeper voices are generally seen as more powerful, competent, and dominant, see D. R. Carney, J. A. Hall, and L. Smith LeBeau, "Beliefs About the Nonverbal Expression of Social Power," *Journal of Nonverbal Behavior* 29 (2005): 106–23, and S. E. Wolff and D. A. Puts, "Vocal Masculinity Is a Robust Dominance Signal in Men," *Behavioral Ecology and Sociobiology* 64 (2010): 1673–83.

41. For studies showing that women with higher-pitched voices are expected to be more physically attractive, see D. R. Feinberg et al., "The Role of Femininity and Averageness of Voice Pitch in Aesthetic Judgments of Women's Voices," *Perception* 37 (2008): 615–23. For the research study showing that women with lower voices are seen as better leaders, see Casey A. Klofstad, Rindy C.

Anderson, and Susan Peters, "Sounds Like a Winner: Voice Pitch Influences Perception of Leadership Capacity in Both Men and Women," *Proceedings of the Royal Society B: Biological Sciences* 279, no. 1738 (2012): 2698–704.

42. Polly Dunbar, "How Laurence Olivier Gave Margaret Thatcher the Voice That Went Down in History," *Daily Mail*, October 29, 2011, www.dailymail.co.uk /news/article-2055214/How-Laurence-Olivier-gave-Margaret-Thatcher-voice -went-history.html.

43. Dana R. Carney, Amy J. Cuddy, and Andy J. Yap, "Power Posing: Brief Nonverbal Displays Affect Neuroendocrine Levels and Risk Tolerance," *Psychological Science* 21, no. 10 (2010): 1363–68.

44. John Brecher, "How to Close the Gender Gap at Work? Strike a Pose," NBC News, January 15, 2014, usnews.nbcnews.com/_news/2014/01/15/22305728 -how-to-close-the-gender-gap-at-work-strike-a-pose.

45. Deborah J. Mitchell, J. Edward Russo, and Nancy Pennington, "Back to the Future: Temporal Perspective in the Explanation of Events," *Journal of Behavioral Decision Making* 2, no. 1 (1989): 25–38.

46. Beth Veinott, Gary A. Klein, and Sterling Wiggins, "Evaluating the Effectiveness of the Premortem Technique on Plan Confidence," in *Proceedings of the 7th International ISCRAM Conference*, Seattle, WA, May 2010.

47. Kimberly A. Daubman, Laurie Heatherington, and Alicia Ahn, "Gender and the Self-Presentation of Academic Achievement," *Sex Roles* 27, nos. 3/4 (1992): 187–204.

48. Laurie Heatherington et al., "Two Investigations of 'Female Modesty' in Achievement Situations," *Sex Roles* 29, nos. 11/12 (1993).

49. The definition of *self-promotion* comes from Alice H. Eagly and Steven J. Karau, "Role Congruity Theory of Prejudice Toward Female Leaders," *Psychological Review* 109, no. 3 (2002): 573–98; quote is on page 584.

50. For a general overview of the research on gender differences in self-promotion, see ibid.

51. Laurie A. Rudman, "Self-Promotion As a Risk Factor for Women: The Costs and Benefits of Counterstereotypical Impression Management," *Journal of Personality and Social Psychology* 74, no. 3 (1998): 629–45. For a more recent review of how women are penalized for exhibiting confidence and drawing attention to their strengths, see Laurie A. Rudman and Julie E. Phelan, "Backlash Effects for Disconfirming Gender Stereotypes in Organizations," *Research in Organizational Behavior* 28 (2008): 61–79.

52. Laurie A. Rudman et al., "Status Incongruity and Backlash Effects: Defending the Gender Hierarchy Motivates Prejudice Against Female Leaders," *Journal of Experimental Social Psychology* 48, no. 1 (2012): 165–79.

53. Sheryl Sandberg writes about her successes and varying feelings about them in *Lean In* (New York: Knopf, 2013). Her reaction to appearing on *Forbes'* annual list of the World's 100 Most Powerful Women can be found on pages 37–38. And Sandberg isn't alone. Research shows that women avoid self-promotion because they're concerned that there will be backlash and retaliation if they promote themselves. See Corinne A. Moss-Racusin and Laurie A. Rudman, "Disruptions in Women's Self-Promotion: The Backlash Avoidance Model," *Psychology of Women Quarterly* 34, no. 2 (2010): 186–202.

54. Jennifer Lawrence, "Why Do I Make Less Than My Male Co-Stars?," *Lenny* (October 13, 2015).

55. Andreas Leibbrandt and John A. List, "Do Women Avoid Salary Negotiations? Evidence from a Large-Scale Natural Field Experiment," *Management Science* 61, no. 9 (2014): 2016–24.

56. Sophie McGovern, "Glove Stretchers and Petticoats: Packing Advice from a Victorian Lady Traveller," GlobetrotterGirls.com, June 11, 2012, accessed June 24, 2015, globetrottergirls.com/2012/06/book-review-hints-for-lady-travellers/.

57. Linda L. Carli, Suzanne J. LaFleur, and Christopher C. Loeber, "Nonverbal Behavior, Gender, and Influence," *Journal of Personality and Social Psychology* 68, no. 6 (1995): 1030–41.

58. Hannah R. Bowles and Linda Babcock, "How Can Women Escape the Compensation Negotiation Dilemma? Relational Accounts Are One Answer," *Psychology of Women Quarterly* 37, no. 1 (2013): 80–96. This scripted language is taken from page 84.

59. Ibid., experiment 2.

60. For a provocative book on unconscious bias, see Mahzarin R. Banaji and Anthony G. Greenwald, *Blindspot: Hidden Biases of Good People* (New York: Delacorte Press, 2013).

5. Stress Makes Her Focused, Not Fragile

1. Dan Bilefsky, "Women Respond to Nobel Laureate's 'Trouble with Girls,'" *New York Times*, June 11, 2015, www.nytimes.com/2015/06/12/world/europe/tim-hunt-nobel-laureate-resigns-sexist-women-female-scientists.html?emc=edit_tnt_20150611&nlid=69372913&tntemailo=y. See also Sarah Knapton, "Sexism Row Scientist Sir Tim Hunt Quits Over 'Trouble with Girls' Speech," *Telegraph*, June 11, 2015, www.telegraph.co.uk/news/science/science-news/11667002/Sexism-row-scientist-Sir-Tim-Hunt-quits-over-trouble-with-girls-speech.html. Deborah Blum, one of the women who spoke at the luncheon and who talked with Hunt afterward to see if he was joking, wrote, "Tim Hunt 'Jokes'

About Women Scientists. Or Not," Storify.com, June 14, 2015, storify.com/deborahblum/tim-hunt-and-his-jokes-about-women-scientists.

2. For the language about breaking down, see Maureen Dowd, "Can Hillary Cry Her Way Back to the White House?," *New York Times,* January 9, 2008, www.nytimes.com/2008/01/09/opinion/o8dowd.html?pagewanted=all&_r=0. See also Jeremy Holden, "Morris, Ingraham Claimed Clinton's Expression of Emotion Raises Questions About Her National Security Credentials," Mediamatters.org, January 8, 2008, mediamatters.org/research/2008/01/08/morris-ingraham-claimed-clintons-expression-of/142089. For "too emotional, too sensitive or too weak," see Emily Friedman, "Can Clinton's Emotions Get the Best of Her?," ABC News, January 7, 2008, abcnews.go.com/Politics/Vote2008/story?id=4097786.

3. Dave Zirin, "Serena Williams and Getting 'Emotional' for Title IX," *Nation,* July 9, 2012, www.thenation.com/blog/168793/serena-williams-and-getting-emotional-title-ix.

4. Donna Britt, "March Sadness: When Male Athletes Turn On the Tears," *Washington Post,* April 3, 2015, www.washingtonpost.com/opinions/march-sadness-when-male-athletes-turn-on-the-tears/2015/04/03/f557c096-d964-11e4-ba28-f2a685dc7f89_story.html.

5. Karen Breslau, "Hillary Clinton's Emotional Moment," *Newsweek,* January 6, 2008, www.newsweek.com/hillary-clintons-emotional-moment-87141.

6. Cathleen Decker, "'Emotional' Dianne Feinstein: At Least She's Not Hysterical," *Chicago Tribune,* April 7, 2014, www.chicagotribune.com/news/politics/la-pn-emotional-dianne-feinstein-cia-20140407,0,6356985.story.

7. John McCain, "Bin Laden's Death and the Debate Over Torture," *Washington Post,* May 11, 2011, www.washingtonpost.com/opinions/bin-ladens-death-and-the-debate-over-torture/2011/05/11/AFd1mdsG_story.html?hpid=z2.

8. Lucy Madison, "Santorum: McCain 'Doesn't Understand' Torture," CBS News, May 17, 2011, www.cbsnews.com/news/santorum-mccain-doesnt-understand-torture.

9. Garrett Quinn, "Analysis: Coakley, Baker Dive into Weeds Again, but Show Their Emotional Sides in Final Debate," MassLive.com, October 29, 2014, www.masslive.com/politics/index.ssf/2014/10/final_debate_analysis.html; Todd Domke, "Baker Wins Debate by 'Losing It,'" WBUR.org, October 29, 2014, www.wbur.org/2014/10/29/domke-debate-baker.

10. Yvonne Abraham, "Turning the Tables," *Boston Globe,* October 30, 2014, www.bostonglobe.com/metro/2014/10/29/abraham/9LxKj9PoVwYdVxhAiVwNhJ/story.html.

11. Agneta H. Fischer, Alice H. Eagly, and Suzanne Oosterwijk, "The Meaning of Tears: Which Sex Seems Emotional Depends on Social Context," *European Journal of Social Psychology* 43 (2013): 505–15.

12. Shauna Shames, "Clearing the Primary Hurdles: Republican Women and the GOP Gender Gap," Political Parity, January 15, 2015, www.politicalparity.org /wp-content/uploads/2015/01/primary-hurdles-full-report.pdf.

13. The title for this section comes from a great paper by Lisa Feldman Barrett and Eliza Bliss-Moreau: "She's Emotional. He's Having a Bad Day: Attributional Explanations for Emotion Stereotypes," *Emotion* 9, no. 5 (2009): 649. Used with permission from Lisa Feldman Barrett (e-mail communication with the author, June 2014).

14. These classic findings were reported in ibid.

15. For research reviewing changes in heart rate and blood pressure, see William Lovallo, "The Cold Pressor Task and Autonomic Function: A Review and Integration," *Psychophysiology* 12, no. 3 (1975): 268–82. For research describing changes in cortisol, see Monika Bullinger et al., "Endocrine Effects of the Cold Pressor Test: Relationships to Subjective Pain Appraisal and Coping," *Psychiatry Research* 12, no. 3 (1984): 227–33.

16. R. van den Bos, M. Harteveld, and H. Stoop, "Stress and Decision-Making in Humans: Performance Is Related to Cortisol Reactivity, Albeit Differently in Men and Women," *Psychoneuroendocrinology* 34, no. 10 (2009): 1449–58.

17. Ibid.

18. S. D. Preston et al., "Effects of Anticipatory Stress on Decision Making in a Gambling Task," *Behavioral Neuroscience* 121, no. 2 (2007): 257.

19. This notion that men strive and compete more than women has been cited in a number of works. I particularly like the blog by Laurence Shatkin, called *Career Laboratory,* in which he mixes "career information and career decision-making in a test tube" to see interesting patterns that arise. In his post on December 9, 2011, titled "Work-Related Values of Men and Women," he draws from the 2003 national survey of college graduates and compares how men and women rank different factors when they are considering a job; see careerlaboratory.blogspot.com/2011/12/work-related-values-of-men-and -women.html.

20. Van den Bos, Harteveld, and Stoop, "Stress and Decision-Making in Humans."

21. It's helpful to note that in the 2009 study where van den Bos and his team looked at high and low responders, they did a median split, which means they took the middle cortisol value and then divided the participants into two groups, those above that value and those below it. It's one effective way to divide a group, but it could be that there's a better division point. In the future,

researchers may set the bar a bit differently if it appears that there's a more precise way to define high and low responders.

22. Andrew J. King et al., "Sex-Differences and Temporal Consistency in Stickleback Fish Boldness," *PLOS ONE* 8, no. 12 (2013): e81116.

23. Jolle Wolter Jolles, Neeltje Boogert, and Ruud van den Bos, "Sex Differences in Risk-Taking and Associative Learning in Rats" (under review by *Royal Society Open Science* in August 2015).

24. Shelley E. Taylor, "Tend and Befriend: Biobehavioral Bases of Affiliation Under Stress," *Current Directions in Psychological Science* 15, no. 6 (2006): 273–77. For her original work comparing the tend-and-befriend reaction to stress with the fight-or-flight approach, see Shelley E. Taylor et al., "Biobehavioral Responses to Stress in Females: Tend-and-Befriend, Not Fight-or-Flight," *Psychological Review* 107, no. 3 (2000): 411–29.

25. Nichole Lighthall and colleagues have used this tend-and-befriend analysis to explain why females may be more risk-alert than risk-seeking. I thank Mara Mather for drawing this interpretation to my attention.

26. I want to credit Ruud van den Bos for making this observation about proactive and reactive coping styles (Ruud van den Bos, personal communication with the author, August 12, 2015). The relationship between prosocial behavior, stress, and gender is, to say the least, a complex one. For a look at how different variables interact, see Tony W. Buchanan and Stephanie D. Preston, "Stress Leads to Prosocial Action in Immediate Need Situations," *Frontiers in Behavioral Neuroscience* 8, no. 5 (2014): 1–6.

27. Christopher Cardoso et al., "Stress-Induced Negative Mood Moderates the Relation Between Oxytocin Administration and Trust: Evidence for the 'Tend-and-Befriend' Response to Stress?," *Psychoneuroendocrinology* 38, no. 11 (2013): 2800–2804. See also R. R. Thompson et al., "Sex-Specific Influences of Vasopressin on Human Social Communication," *Proceedings of the National Academy of Sciences* 103, no. 20 (2006): 7889–94.

28. L. Tomova et al., "Is Stress Affecting Our Ability to Tune Into Others? Evidence for Gender Differences in the Effects of Stress on Self-Other Distinction," *Psychoneuroendocrinology* 43 (2014): 95–104.

29. Ruud van den Bos shared this metaphor with me in a personal communication in June 2014. He also uses this metaphor in this article: Ruud van den Bos, Jolle W. Jolles, and J. R. Homberg, "Social Modulation of Decision-Making: A Cross-Species Review," *Frontiers in Human Neuroscience* 7 (2013): 301.

30. Companies with no female board representation experienced only 10 percent growth from 2005 to 2011, whereas companies with at least one woman on the board enjoyed 14 percent growth during this same difficult period, a 40

percent improvement over their peers; see Urs Rohner and Brady W. Dougan, eds., "Gender Diversity and Corporate Performance," Credit Suisse AG, August 2012, 14.

31. Ibid.

32. For companies with large-cap stocks, over $10 billion, those corporations that had at least one woman on the board outperformed corporations with all-male boards. The advantages of having a woman in the boardroom were slightly less for smaller companies. For companies with small-to-mid-cap stocks, defined as less than $10 billion a year, companies with at least one woman on the board outperformed companies with all-male boards by 17 percent. See figures 9 and 10 in ibid.

33. For the meta-analysis that finds a country's gender parity appears to be a moderator of the relation between board diversity and market performance, see Corinne Post and Kris Byron, "Women on Boards and Firm Financial Performance: A Meta-Analysis," *Academy of Management Journal* 58, no. 5 (2015): 1546–71. A fantastic paper summarizing and scrutinizing the contradictory findings on firm performance and the presence of female board members was written by Alice H. Eagly, "When Passionate Advocates Meet Research on Diversity: Does the Honest Broker Stand a Chance?," *Journal of Social Issues* (in press). It's also not clear what comes first. Perhaps the sharpest organizations do everything right: they outperform the market and promote more women. Perhaps the most profitable companies have more money to hire and retain those women with the savviest insights, an argument made by Eagly in her paper.

34. Estimated costs of the U.S. government shutdown are taken from Steve James, "Money for Nothing: Government Shutdown Costs $12.5 Million per Hour," NBC News, October 1, 2013. The number of federal employees furloughed is taken from Laura Meckler and Rebecca Ballhaus, "More Than 800,000 Federal Workers Are Furloughed," *Wall Street Journal,* October 1, 2013, online.wsj.com /news/articles/SB10001424052702304373104579107480729687014.

35. *2013 Catalyst Census: Fortune 500 Women Board Directors* (December 10, 2013), http://www.catalyst.org/knowledge/2013-catalyst-census-fortune-500-women -board-directors.

36. Susan Vinnicombe, Elena Dolder, and Caroline Turner, *The Female FTSE Board Report 2014* (Bedford, UK: Cranfield School of Management, Cranfield University, 2014).

37. Ryan and Haslam technically weren't the first to notice this pattern. Elizabeth Judge, in 2003, wrote an article called "Women on Board: Help or Hindrance?" for the *Times* that made Ryan and Haslam curious. Judge noted that there was

a disturbing pattern among the top one hundred companies on the London Stock Exchange — namely, that companies with women directors weren't performing as well as companies with male directors. Judge thought the female leaders were to blame, but Ryan and Haslam looked further into the issue and discovered that the companies with low stock performance had been struggling, on average, five months before the women were appointed. So women weren't to blame for the slide in performance. They had been brought in long after the slide had begun.

38. Sylvia Maxfield, "Janet Yellen on the Glass Cliff," *PC News,* November 4, 2013, www.providence.edu/news/headlines/Pages/Sylvia-Maxfield-Providence -Journal-op-ed.aspx.

39. Catherine Fox, "Lagarde Appointment Tests 'Glass-Cliff' Theory," *Financial Review,* July 26, 2011, www.afr.com/p/opinion/lagarde_appointment_tests _glass_DLVt4mTpAqOTaxhjxL8y8H.

40. Jaclyn Trop, "Is Mary Barra Standing on a Glass Cliff?," *New Yorker,* April 30, 2014. Regarding the cover-up, see Rebecca Ruiz and Danielle Ivory, "Documents Show General Motors Kept Silent on Fatal Crashes," *New York Times,* July 15, 2014.

41. Nichole R. Lighthall et al., "Gender Differences in Reward-Related Decision Processing Under Stress," *Social Cognitive and Affective Neuroscience 7,* no. 4 (2012): 476–84. Other researchers have similarly found that under stressful conditions, both men and women attend more to positive feedback than negative feedback when they're learning something new; see Antje Petzold et al., "Stress Reduces Use of Negative Feedback in a Feedback-Based Learning Task," *Behavioral Neuroscience* 124, no. 2 (2010): 248.

42. Paul C. Nutt, "The Identification of Solution Ideas During Organizational Decision Making," *Management Science* 39 (1993): 1071–85. If you want to learn more about the many problems with "whether or not" and "should I or shouldn't I" decisions, see a later paper: Paul C. Nutt, "Surprising but True: Half the Decisions in Organizations Fail," *Academy of Management Executive* 13, no. 7 (1999): 5–90.

43. Hans Georg Gemuden and Jurgen Hauschildt, "Number of Alternatives and Efficiency in Different Types of Top-Management Decisions," *European Journal of Operational Research* 22 (1985): 178–90. It might sound a bit discouraging that so few of the firm's decisions were deemed "very good," but the firm was evaluating its own past decisions and the team had very high standards. They rated 34 percent of all decisions across an eighteen-month period as poor, 40 percent as satisfactory, and only 26 percent as very good.

44. Barry Schwartz, *The Paradox of Choice* (New York: Ecco, 2004). The book is

considered a classic by decision-making researchers, but there's a lively debate today about choice overload and when it occurs and when it does not. For a meta-analysis of fifty studies that tested whether choice overload occurs, see Benjamin Scheibehenne, Rainer Greifeneder, and Peter M. Todd, "Can There Ever Be Too Many Options? A Meta-Analytic Review of Choice Overload," *Journal of Consumer Research* 37, no. 3 (2010): 409–25. They found that the results vary widely from study to study, some studies noting that people have reduced motivation and satisfaction when they have too many choices, and others finding no change whatsoever in motivation or satisfaction. Part of it depends on the expertise of the person doing the choosing. If you love tools, then standing in front of a wall of screwdrivers would bring a smile to your face. Chip and Dan Heath do a fantastic job analyzing the practical implication of choice overload research in their book *Decisive: How to Make Better Decisions in Life and Work* (New York: Crown Business, 2013).

45. Alison Wood Brooks, "Get Excited: Reappraising Pre-Performance Anxiety as Excitement," *Journal of Experimental Psychology: General* 143, no. 3 (2014): 1144–58.

46. Ibid.

47. Ibid. See also the classic work by S. Schachter and J. Singer, "Cognitive, Social, and Physiological Determinants of Emotional State," *Psychological Review* 69 (1962): 379–99, doi:10.1037/h0046234.

48. In case you're wondering, she picked that song because it's so well known by English speakers, the twenty-first-most-downloaded song in iTunes history. It's also possible to sound good in three different octaves, so women with high voices could perform as well as men with low ones.

49. Brooks, "Get Excited," experiments 2 and 3.

50. Jeremy P. Jamieson, Matthew K. Nock, and Wendy Berry Mendes, "Mind Over Matter: Reappraising Arousal Improves Cardiovascular and Cognitive Responses to Stress," *Journal of Experimental Psychology: General* 141, no. 3 (2012): 417–22.

6. Watching Other People Make Terrible Decisions

1. Ulrica G. Nilsson et al., "The Desire for Involvement in Healthcare, Anxiety and Coping in Patients and Their Partners After a Myocardial Infarction," *European Journal of Cardiovascular Nursing* 12, no. 5 (2013): 461–67, doi:10.1177/1474515112472269.

2. Linda J. Sax, Alyssa N. Bryant, and Casandra E. Harper, "The Differential Effects of Student-Faculty Interaction on College Outcomes for Women and Men," *Journal of College Student Development* 46, no. 6 (2005): 642–57.

3. Yeonjung Lee and Fengyan Tang, "More Caregiving, Less Working: Caregiving Roles and Gender Difference," *Journal of Applied Gerontology* 34, no. 4 (June 2015), doi:10.1177/0733464813508649.

4. Francesca Gino describes this research in her book *Sidetracked: Why Our Decisions Get Derailed and How We Can Stick to the Plan* (Boston: Harvard Business Review Press, 2013). The full research study and data are in Francesca Gino, "Do We Listen to Advice Just Because We Paid for It? The Impact of Advice Cost on Its Use," *Organizational Behavior and Human Decision Processes* 107, no. 2 (2008): 234–45.

5. Paul C. Nutt, "The Identification of Solution Ideas During Organizational Decision Making," *Management Science* 39 (1993): 1071–85. Also see chapter 2, "Avoid a Narrow Frame," in Chip and Dan Heath, *Decisive: How to Make Better Decisions in Life and Work* (New York: Crown Business, 2013).

6. This example is taken from Daniel Kahneman, *Thinking, Fast and Slow* (New York: Farrar, Straus and Giroux, 2011), 279–80.

7. Daniel Kahneman and Amos Tversky, "Prospect Theory: An Analysis of Decision Under Risk," *Econometrica: Journal of the Econometric Society* (1979): 263–91. Kahneman and Tversky developed a new version of the theory over a decade later; see Amos Tversky and Daniel Kahneman, "Advances in Prospect Theory: Cumulative Representation of Uncertainty," *Journal of Risk and Uncertainty* 5, no. 4 (1992): 297–323.

8. Daniel Kahneman, personal communication with the author, May 5, 2015.

9. Elanor F. Williams and Robyn A. LeBoeuf, "Starting Your Diet Tomorrow: People Believe They Will Have More Control Over the Future Than They Did Over the Past," *Journal of Consumer Research* (manuscript under revision). In some of the experiments that Williams and LeBoeuf conducted, they asked people about lived experiences, and in some of the studies, they asked participants to imagine hypothetical situations.

10. Ironically, or perhaps quite realistically, Williams and LeBoeuf found that one of the few places where people doubted they would have more control in the future was in how much they would procrastinate. The thinking seems to be, *Once a dawdler, always a dawdler.* People probably have more control over how much they procrastinate than over a game of chance, but they see procrastination as something they can't control.

11. Daniel T. Gilbert et al., "Looking Forward to Looking Backward: The Misprediction of Regret," *Psychological Science* 15, no. 5 (2004): 346–50.

12. Kriti Jain et al., "Unpacking the Future: A Nudge Toward Wider Subjective Confidence Intervals," *Management Science* 59, no. 9 (2013): 1970–87. Admittedly, I'm applying the research on unpacking the future to different

kinds of problems, to social and relationship problems. Most of the research on unpacking has addressed how people are poor at estimating how much time or control they will have on a task they need to perform in the future. For more on unpacking the future and how it improves decision-making, see Jack B. Soll, Katherine L. Milkman, and John W. Payne, "A User's Guide to Debiasing," in *Wiley-Blackwell Handbook of Judgment and Decision-Making,* eds. Gideon Keren and George Wu (Malden, MA: Wiley-Blackwell, 2015).

13. I want to extend my gratitude to Nora Williams for helping me generate these strategies to address the bias of believing we'll have more control in the future. Like the generous colleague she is, she offered several clever ideas.

14. If you want to learn more about cognitive dissonance, the most delightful and insightful book I've read on the topic is Carol Tavris and Elliot Aronson, *Mistakes Were Made (but Not by Me),* rev. ed. (Boston: Houghton Mifflin Harcourt, 2015). The examples are taken from recent news stories and recent history, and Tavris and Aronson write well.

15. Leon Festinger wrote much about cognitive dissonance over the years, but he laid out his complete theory in his book *A Theory of Cognitive Dissonance* (Stanford, CA: Stanford University Press, 1957).

16. It may sound outrageous to suggest that parents heavily edit or even write their children's college application essays, but it's a trend that college and university admissions offices have noticed and decry. See Rebecca Joseph, "A Plea to Those Helping Students with College Application Essays: Let the 17-Year-Old Voice Take Center Stage," *Huffington Post,* October 17, 2013. See also Kevin McMullin, "For parents: No essay hijacking," CollegeWise.com, accessed January 19, 2016, www.collegewise.com/for-parents-no-essay-hijacking.

17. Tavris and Aronson point out in their book that this is what we're really saying when we ask, "What were you thinking?"

18. Anthony Pratkanis and Doug Shadel, *Weapons of Fraud: A Course Book for Fraud Fighters* (Washington, DC: AARP, 2005). To request a free copy of this book, go to www.aarp.org.

19. This definition of the positivity effect is taken from the paper by Andrew E. Reed and Laura L. Carstensen, "The Theory Behind the Age-Related Positivity Effect," *Frontiers in Psychology* 3 (2012): 1–9.

20. Derek M. Isaacowitz et al., "Selective Preference in Visual Fixation Away from Negative Images in Old Age? An Eye-Tracking Study," *Psychology and Aging* 21, no. 1 (2006): 40–48.

21. Mara Mather, Marisa Knight, and Michael McCaffrey, "The Allure of the Alignable: Younger and Older Adults' False Memories of Choice Features," *Journal of Experimental Psychology: General* 134, no. 1 (2005): 38–51.

22. Sunghan Kim et al., "Age Differences in Choice Satisfaction: A Positivity Effect in Decision Making," *Psychology and Aging* 23, no. 1 (2008): 33.

23. Corinna E. Löckenhoff and Laura L. Carstensen, "Aging, Emotion, and Health-Related Decision Strategies: Motivational Manipulations Can Reduce Age Differences," *Psychology and Aging* 22, no. 1 (2007): 134–46. See also Corinna E. Löckenhoff and Laura L. Carstensen, "Decision Strategies in Health Care Choices for Self and Others: Older but Not Younger Adults Make Adjustments for the Age of the Decision Target," *Journals of Gerontology Series B: Psychological Sciences and Social Sciences* 63, no. 2 (2008): 106–9.

24. Mara Mather and Laura L. Carstensen, "Aging and Motivated Cognition: The Positivity Effect in Attention and Memory," *Trends in Cognitive Sciences* 9, no. 10 (2005): 496–502. See also Laura L. Carstensen and Joseph A. Mikels, "At the Intersection of Emotion and Cognition: Aging and the Positivity Effect," *Current Directions in Psychological Science* 14, no. 3 (2005): 117–21.

25. Quinn Kennedy, Mara Mather, and Laura L. Carstensen, "The Role of Motivation in the Age-Related Positivity Effect in Autobiographical Memory," *Psychological Science* 15, no. 3 (2004): 208–14.

26. You might be wondering if these were exceptionally healthy nuns, with all of them alive and well fourteen years after the original survey. Unfortunately, many of the nuns had passed away. Of the 862 nuns who had completed the survey in 1987, 316 were still alive in 2001, and 300 of them participated in the second survey. The original study done in 1987 was unpublished but has been cited as Laura L. Carstensen and K. Burrus, "Stress, Health and the Life Course in a Midwestern Religious Community" (unpublished manuscript, 1996). You might also be wondering why the researchers chose to study nuns, of all people. I asked Mara Mather, one of the authors on the follow-up study, and she explained it wasn't that they thought the nuns would have a more or less positive outlook on the world around, it just simplified the methodology. Many people move around or get married and change names, making them hard for researchers to track down after a decade or more has passed. It's much easier to study members of a religious community over fourteen years than it is to study average members of the population. Other researchers have studied nuns in longitudinal studies of Alzheimer's, weight change, and blood pressure.

27. For a review of how the positivity effect affects the memories that inform our decisions, see Quinn Kennedy and Mara Mather, "Aging, Affect, and Decision Making," in *Do Emotions Help or Hurt Decision Making?*, eds. K. Vohs, R. Baumeister, and G. Loewenstein (New York: Russell Sage Foundation, 2007), 245–65.

28. For research on how the positivity effect could be a sign of strong emotional

and mental health, see Laura K. Sasse et al., "Selective Control of Attention Supports the Positivity Effect in Aging," *PLOS ONE* 9, no. 8 (2014), e104180. For research on how controlling one's emotions requires an interactive network of brain areas, see Michiko Sakaki, Lin Nga, and Mara Mather, "Amygdala Functional Connectivity with Medial Prefrontal Cortex at Rest Predicts the Positivity Effect in Older Adults' Memory," *Journal of Cognitive Neuroscience* 25, no. 8 (2013): 1206–24.

29. Löckenhoff and Carstensen, "Decision Strategies in Health Care Choices."

30. Kennedy, Mather, and Carstensen, "The Role of Motivation."

31. For more on emotional regulation and aging, see Reed and Carstensen, "The Theory Behind the Age-Related Positivity Effect," and M. Mather, "The Emotion Paradox in the Aging Brain," *Annals of the New York Academy of Sciences* 1251, no. 1 (2012): 33–49.

32. James J. Gross et al., "Emotion and Aging: Experience, Expression, and Control," *Psychology and Aging* 12, no. 4 (1997): 590. For a more recent study, see Laura L. Carstensen et al., "Emotional Experience Improves with Age: Evidence Based on Over 10 Years of Experience Sampling," *Psychology and Aging* 26, no. 1 (2011): 21–33.

33. For a review of how older adults seek fewer choices, see R. Mata and L. Nunes, "When Less Is Enough: Cognitive Aging, Information Search, and Decision Quality in Consumer Choice," *Psychology of Aging* 25 (2010): 289–98, and Mara Mather, "A Review of Decision Making Processes: Weighing the Risks and Benefits of Aging," in *When I'm 64,* eds. L. L. Carstensen and C. R. Hartel (Washington, DC: National Academies Press, 2006), 145–73. For a recent neuroscientific study examining why older adults seek less information in their decision-making, see Julia Spaniol and Pete Wegier, "Decisions from Experience: Adaptive Information Search and Choice in Younger and Older Adults," *Frontiers in Neuroscience* 6 (2012).

34. Andrew E. Reed, Joseph A. Mikels, and Kosali I. Simon, "Older Adults Prefer Less Choice Than Young Adults," *Psychology and Aging* 23, no. 3 (2008): 671–75. For research on how older women seek less information about breast cancer treatment than younger women, see Bonnie Meyer, Connie Russo, and Andrew Talbot, "Discourse Comprehension and Problem Solving: Decisions About the Treatment of Breast Cancer by Women Across the Life Span," *Psychology and Aging* 10, no. 1 (1995): 84.

35. Laura L. Carstensen et al., "Emotional Experience Improves with Age," 21.

36. Marisa Knight et al., "Aging and Goal-Directed Emotional Attention: Distraction Reverses Emotional Biases," *Emotion* 7, no. 4 (2007): 705. See also M. Mather and M. Knight, "Goal Directed Memory: The Role of

Cognitive Control in Older Adults' Emotional Memory," *Psychology and Aging* 20, no. 4 (2005): 554–70. For work challenging the claim that distraction reduces the positivity effect in the elderly, see E. S. Allard and D. M. Isaacowitz, "Are Preferences in Emotional Processing Affected by Distraction? Examining the Age-Related Positivity Effect in Visual Fixation Within a Dual-Task Paradigm," *Aging, Neuropsychology, and Cognition* 15, no. 6 (2008): 725–43.

Afterword

1. Elizabeth Loftus is an expert on the fallibility of eyewitness testimony; see Elizabeth Loftus, "Our Changeable Memories: Legal and Practical Implications," *Nature Reviews Neuroscience* 4, no. 3 (2003): 231–34.

2. I adapted this painting analogy of memory from a philosophy forum thread titled "What's a Good Analogy for Human Memory?," posted October 17, 2012. I thank user MyselfYourself, whoever you are, for this clever idea; see forums.philosophyforums.com/threads/whats-a-good-analogy-for-human -memory-57001.html.

3. For research on voters updating their political beliefs following an election, see Ryan K. Beasley and Mark R. Joslyn, "Cognitive Dissonance and Post-Decision Attitude Change in Six Presidential Elections," *Political Psychology* 22, no. 3 (2001): 521–40, and Linda J. Levine, "Reconstructing Memory for Emotions," *Journal of Experimental Psychology: General* 126, no. 2 (1997): 165–77. For research on adults misremembering their high-school grades, see Harry P. Bahrick, Lynda K. Hall, and Stephanie A. Berger, "Accuracy and Distortion in Memory for High School Grades," *Psychological Science* 7, no. 5 (1996): 265–71. For research on adults misremembering their teenage personalities, see Daniel Offer et al., "The Altering of Reported Experiences," *Journal of the American Academy of Child and Adolescent Psychiatry* 39, no. 6 (2000): 735–42.

4. Carol Tavris and Elliot Aronson, *Mistakes Were Made (but Not by Me)*, rev. ed. (Boston: Houghton Mifflin Harcourt, 2015), 101.

Acknowledgments

1. Michael Chabon is a Pulitzer Prize–winning novelist, and the full quote is this: "You need three things to become a successful novelist: talent, luck, and discipline. Discipline is the one element of those three things that you can control, so that is the one you have to focus on controlling, and you just have to hope and trust in the other two."

Index